An Ignoble End

Jacqueline James M.P.

AuthorHouse™ UK Ltd.
500 Avebury Boulevard
Central Milton Keynes, MK9 2BE
www.authorhouse.co.uk
Phone: 08001974150

First published by AuthorHouse 12/3/2009

ISBN: 978-1-4490-1242-7 (sc)

This novel is a work of fiction. Names, Characters, locations, or Political Parties, are the product of the author's imagination. Any resemblance to persons,living or dead, is entirely coincidental. Some medical and parliamentary details have been altered for the sake of telling a good story.

This book is printed on acid-free paper.

SYNOPSIS

It's a miserable rainy day in late 1995. The discovery of a malignant tumour in her right breast is the beginning of a ten year roller-coaster ride of despair, bravery, determination and elation. Back in her biology teaching job, after successful mastectomy and chemotherapy, all was well with the world, and she was more than happy. Then, out of the blue, a new opportunity presented itself, by way of a local election candidate calling to canvass her political leanings She was so impressed, she joined the local party and became more involved in politics, at a local level to begin with. She took to it like a duck to water, and did very well as a ward councillor, and began to think about a career in main stream politics.With her grit and determination she duly won the Parliamentary seat for her own constituency. The early days of her parliamentary career were something of a shock. The one thing she and her husband had not properly considered was the fact that they were apart for four days a week, but a problem that was easily overcome when she employed him as her Parliamentary researcher. With her dream job, living the high life in London, what could be better. Till one day, the bravery and determination of yesteryear were to desert her in the wake of a second malignant tumour, in her spine. As it was deemed inoperable, and as a consequence, life ending. All that is left is despair, and determination. Despair and determination to wrest the best possible deal out of this agonizing situation, not for herself, but her husband and twin sons. To resign her seat, and get nothing would be a bitter blow. The trick is, to be a member of Parliament when she dies. The second half of the story is political roller-coaster ride of lies, fraud , deception. Culminating in perjury of the worst kind.

AN IGNOBLE END

As we pulled into the gravelly, oily, water filled pot-holed car park, we sat quietly for a moment as we watched the rain drizzle down the wind screen. Neither of us could think of anything much to say, there was nothing much to say, just sit and watch the world go by in a wet spray at the end of the road. We were a bit early, I suppose we just wanted to get this day over with, a day that was destined to change my life forever. As we stepped from the car we had to hop and skip over and around the puddles, I was so nervous I nearly tripped as we approached the giant oak doors, with a carved stone lintel telling us it was built by William Dorcass in 1901. Doors that have guarded the entrance to this imposing red brick former work house for the best part of a hundred years. Once inside, we were met by the most uninspiring sight of grey walls, grey plastic tiles covering the floor in a grey painted corridor on a grey day, in early September 1995 With my heart pounding like a drum I couldn't link my husband's arm as I usually do when we go out together for fear he would feel me trembling like a nervous kitten in a strange new home. Not a very good feeling whilst waiting in that little green painted waiting room with very little heat from those cast iron radiators, the out of date magazines, and odd chairs. This poor little waiting looked just as miserable as I felt. We are hear to be told weather or not our world was about to fall apart. The results of my biopsy were waiting for us. This was potentially the worst day of my not very long life of forty seven years. We are here because when in the shower a few weeks ago, I felt a small lump in my breast, as I realised what this could mean, I began to think the worst. The lump was just under my right nipple, probably that's why I didn't feel it sooner. By the time I had dried myself and put on my bath robe I was in one hell of a state, so I went downstairs to put the kettle on, "there's a lot of brewing up in this story". My first thought was, how was I going to tell my husband, what about the boys, do I tell them today or do I wait, as luck had it, they were both away at university, so that was something to be thank full for. I knew it would be folly to put things off, if only for a week or two, so I bit the bullet and phoned my G.P. Even though the receptionist could tell I was very worried, she wouldn't budge and said I should try again tomorrow because they were so busy. I put the phone down, sank to the floor and sobbed my heart out. It wasn't long before I heard my husband's car coming up the drive, I knew I had only a few minutes to compose myself as he put the car in the garage. He came through the door with his usual happy smile. As I looked at him I felt so sorry, I was about to bring his world crashing in. He really looked forward to these two precious days to wind down. As usual we sat at the kitchen table with a cup of tea, then as he looked at me, he realised I'd been crying, in his lovely soft voice he said to me, what ever is the matter, why the tears. As he sipped his tea, I told him about my discovery and that I had already phoned the doctor for an appointment and that the receptionist was very terse, she said even if the doctor saw me today, there wasn't much anyone could do till Monday in any case. Even when I told her what I thought the problem was, she told me to try again on Monday morning. Needless to say my husband was furious with my treatment at such a difficult time, and said he would go to the surgery first thing tomorrow morning. He asked if I had 'phoned my mother, I said I was going to, then thought better of it. I'll wait till I know something more definite, not much point worrying the poor woman particularly as she lives in Kendal in the lake district. She and my dad worked for 'K' shoes for donkeys years, even though she's got all her faculties she's very frail. We did ask her several times if she would like to come and live with us since dad died a couple of years ago. To say he died isn't quite right, he was hit by a bus on Lound road near the 'K'.shoes factory one horrible snowy January evening in 1993. Come to think of it, maybe it's not a bad way to go, but not yet ah. I did ring my mother, as I do every week, but I didn't say anything about my lump. Even though I rang her every week, I could never find the right time to tell her, till I thought why dose she ever need to know. It's funny really, well, not exactly funny, but an odd quirk of fete, neither my husband or I had brothers or sisters, and each of us have lost parents. Now we both have to deal with this demon. I couldn't prepare anything for tea, I was so worried, it's just as well, because I couldn't eat a thing anyway, nor could my husband, that's most unusual for him, he has such a good appetite, well, we had just been knocked for six, but a cup of tea always seems to help. After a bit of tele', I couldn't concentrate at all, a couple or three glasses of wine, and some very unappetising supper, we decided to have an early night, no! not that sort of an early night, but we did snuggle up together, and surprisingly enough, had a good nights sleep, must have been the wine. Early next morning, about eight thirty, my husband brought me a lovely hot cup of tea and sat on the edge of the bed for a while. Then he said he was off to the doctors to give them a piece of his mind, but I know he's far too nice to do such a thing. I wished him luck as he went down the stairs, don't think I'll be too long he shouted as he closed the front door. That was my cue to get my backside into gear, have a shower, having a shower now and being to scared to look at myself to close, puts me in mind of the shower scene in that Hitchcock film, Psycho. It may sound silly, but I had a very quick shower, and made myself look more like the woman he married all those years ago. I'm glad I did, I felt a lot better, better than the snivelling wreck of a woman of last night. I finished making myself look a little more presentable minutes before he came back with a look of satisfaction on his face. He told me the receptionist apologised for yesterday, and said she was very busy. She was so much more sympathetic this morning, and made an appointment for me to see my doctor on Monday at nine thirty. Then he took a deep breath, looked at me with his big brown eyes, and gave me a big hug and told me how much he loves me, but was unusually quiet over the weekend. We went to the supermarket as usual, people say never shop when your hungry, it was easy for us, neither of us had much of an appetite at the moment.

We just bought a few essentials milk, bread veg' you know the sort of things. On our way home we had a look round the local garden centre, he bought himself some new wellies' for the winter, and I bought some strong gloves, strong gloves was the last thing I needed at the moment, more like a strong drink, I just bought them for something to do. I couldn't help but think there was something not quite right, I couldn't quite understand why he was so quiet. Then, on our way home, it hit me like a tonne of bricks! The poor man, no wonder he was so quiet. When he was in his teens, he watched his father nurse his poor mother through years of sickness of one kind or another, culminating in cancer of the bowl and intestine. She died when he was only twenty years old. His father died a few years ago. I can't remember how he died, and I sure ain't going to ask him, now I was unusually quiet. When we got home, he tried on his new wellies, like you do. I put the kettle on and began to put the groceries away, and asked him what he wanted for tea, and tried to lighten the atmosphere as I admired his wellies. But it was no use, I couldn't let him suffer his memories in silence. I told him I could understand his anguish at a time like this, I knelt down and put my head on his lap and just cried for him. He realised why I was crying, and did his best to comfort me, saying it was along time ago. Be that as it may, I couldn't help having that feeling of despair, not least because our sons may well have to watch me suffer a similar fate. We both agreed to try not to let it spoil the rest of the weekend and tried to get on with life and to make the best of what we have. Sunday was fine but a bit cool, my husband just had to try out his new wellies as he swept the half a tonne of leaves from the front garden alone, put them in plastic bags and went to the local waste disposal site. He was happy now he'd tried out his wellies, like a big kid. After tea we both had a little doze in the chair, must be all that fresh air, and dumplings we had. He took Monday morning off work. He said Monday is usually a bit of a mad house, but they'll just have to manage without me won't they, as neither he, nor I, wanted me to go to see the doctor alone. As we sat there with a dozen or so people in the waiting room, I began to think, it's funny, but no one looks as if they have anything wrong with them do they, I certainly looked the picture of health. As we waited to see the doctor, my nerves were at breaking point. My husband did his best to reassure me as he took hold of my hand, saying that my lump could well be benign, but I was still worried sick, and thought it's either one or t'other and could just as easily be malignant, and of course the doctor had no way of telling, even if she was fairly sure, she couldn't say anything difinitive, just in case she was wrong. My visit hear, was only to confirm what I already knew, this is just the first step on what will become a long journey, a journey of hope, relief, optimism, despair and, well, you'll just have to read on. I don't know why, but I asked my husband to wait for me in the waiting room, but he said he would get a breath of air for a few minutes as I went in to hear what the doctor had to say. I'm not going to give you any detail about my examination, my husband wasn't there, so why should I tell you. Suffice to say she did confirm my own diagnosis. I did have a lump under my right breast, and I wasn't growing a third nipple. On our way home, in our new car, a Rover 620 ti, very posh, well, he works hard, but we couldn't enjoy it as we sat in those lovely brown leather seats, with the engine as quiet as a purring kitten. I didn't say a word, and thought I'll wait till we get home. It's a good job he's a patient man, he didn't press me at all.

As we sat at the kitchen table with our usual cup of tea, I explained what the situation was. I told him the doctor did confirm I had a lump under my right breast, and I would need further examinations, but she wouldn't be pushed into saying what she thought it could be, she just said we'll have to wait and see. My husband said impatiently, has she made an appointment for more tests? I held out my hand to reveal a screwed up card, and said she would arrange for me to have a diagnostic mammogram at the local out patients clinic one day next week, next week! he said, why not sooner! and here's me telling you he's a patient man. About an hour later the 'phone rang, my husband answered it, and after a few hums and hars, as he wrote something down on the note pad by the 'phone, he said thank you and replaced the receiver. He said the doctor had just been onto the clinic, they will send a letter with details of my appointment and expect to see me next Monday, at nine o'clock. My husband was not best pleased with such a delay, but there wasn't much we could do about it at this stage. Someone once said a week is a long time in politics. Well, they should try waiting for the results of these tests. All my husband could say was, that we'll fight this together, every step of the way. I'm glad he was so positive, as I had the feeling he had put his own tragedy as a young man behind him and was going to be as strong as he could for me and the boys. Whom?, by the way, don't know anything yet. A week really is a long time to wait, so I kept myself busy with work and tried to get on with my life as best I could. We did get there in the end. This time

I had to go to the clinic on my own, as my husband had to work on some sort of rush job. He didn't want me to go alone, and rather foolishly said could we book another appointment, or see if there are any cancellations. I must say I blew my top and said I'm going for a mammogram, not bloody hair do! Then I assured him I would be o.k. as much as I wanted him with me, I said I've been waiting and worrying all week, and just wanted to get it over with. I didn't mean to raise my voice to him, but, on this occasion, I did. My husband wanted to give me a lift, but couldn't be late for his meeting at nine o' clock. I was going to ring for a taxi, then thought, I didn't want the taxi driver knowing where I was going, silly really, but there you are. As I strolled down the road to the bus stop I couldn't help but think of my husband trying to concentrate, and wishing , as I was, that he was with me. Then I thought, he probably is, in his thoughts. A mammogram is a simple procedure really, a sort of squashed x-ray of both breasts. They always ex-ray both breasts, just to be sure. And of course there's the ubiquitous form filling. The nurse asked me to fill in a form as long as your arm. Any history of cancer in the family, my menstrual cycle, birth control, child bearing, nipple discharge, breast pain, mother's maiden name, well not really, but you get the picture. As I got dressed and waited for the nurse to tell me what my next move would be, she gave me a card just to say that I had had my mammogram on that day and told me I could expect a call from my G.P. in due course, and I could go home. I was in and out in under an hour.

It was my intention to go back to work in the afternoon, they didn't know anything about my mammogram, I just said I was going to the doctors. But as it was such a nice day I went into town to look at the shops, and get some fresh fish for our tea, cod fillet if I can, from the guy in the market. I arrived home about two o'clock. The first thing I did was to ring my husband, no, sorry, that was the second. I put the kettle on first, then I rang him. He asked me how it went and said I'd been a long time. When I told him I was at home he thought there was something wrong, no no no! I'm fine and said I had a walk round town as it was such a nice day. Good for you he said. Then he asked some silly questions, like, did it hurt, was it embarrassing, was the mammographer a man or woman, don't be silly I said, I'm going into the garden shortly, so buzz off, I'll see you tonight. Of course I had to give my him chapter and verse when he came home from work. The fish with some new Cypress potatoes was beautiful by the way. Once again we had to wait over a week for the results. I couldn't help but think the weeks soon add up, and is more than long enough to have to wait, but we had little choice. I've been so wrapped up in my story, I forgot to tell you what my job is. This is going to sound a bit big headed, but well, it's true. I'm a clever sod really, I teach biology at a posh girls school, just a short bus ride away. Being a biology teacher dose have it's down side, inasmuch as I have a pretty good idea of what this lump is, something I kept to myself, thinking of my husband's teenage trauma, I still have cause for concern, that's why I'm so worried, not just for me, but my husband and our two sons. These are the father's footsteps we don't want them to follow in, watching their mum suffering like this, as their dad must have done all those years ago. My teaching job gives me plenty to think about and helped me to put my worries to the back of my mind, for a while at least. Once again we were off to see my doctor, with growing trepidation as to what the future may hold in store for us. As we waited to see her about the results of my mammogram my husband tried to put my mind at ease by saying they might be wrong, and in any case, the lump could be benign. But it was useless, I had convinced myself it was breast cancer. It's getting on for three weeks of waiting and worrying since that first 'phone call, but that was nothing compared to this final twenty minute wait to see my doctor for the results of my mammogram, well not results exactly, just one of the many steps on this very long journey. It was the longest twenty minutes ever as we sat there in the hushed waiting room. My mind was in turmoil, is it really a tumour? if it is, is it malignant? is it benign? was I going to die before I had started to live? I was just beginning to brake down, when the receptionist called out my name to see the doctor. We both felt that same sinking feeling, with the knots in the pit of the stomach one gets with the expectation of bad news. As we sat down, I straitened my back and put my hands on the edge of her desk to stop myself trembling, she could see just how I was feeling and did her best to reassure me, but no matter how she tried to wrap it up with bells and bows, it was still bad news as she told me the worst news any woman could ever wish to here as she confirmed that I did have a tumour in my right breast, but that was not all, she took hold of my hands, looked at my husband, then back at me and said the mammogram found an abnormality in my left breast as well. It was all I could do not to break down then and there, I just couldn't hold it together and had to get out of there and leave my poor husband to pick up the pieces. After a few minutes he came over to me at the far end of the car park, he just held out his arms as he walked towards me. What a comfort he was at times like this, I really don't deserve such a man. As we drove home I could still hear the doctor's words as she tried to reassure me by saying the lump in my right breast could be benign, and the anomalies in my left breast, equally, could be just an anomaly, but I knew the odds were against it. I thought my whole world had collapsed, or to be more precise, our world. How were we going to break the news to the boys? I was on the verge of tears as my husband squeezed my hand tightly. He told me the doctor said I would need to have a biopsy as soon as possible. Sounds a bit urgent I said. He went on to say the doctor told him, once a person has had a mammogram, showing anomalies they are all treated as urgent. The doctor made an appointment for the next afternoon at the brand new cancer hospital a few miles away. She gave my husband a letter of introduction for me to take, wished us well and said try not to worry.

The church clock struck one as we walked through the modern glass and polished steel entrance into the unknown, but an entrance that was to become all to familiar over the next few months. We sat in comfortable high backed chairs in a really modern warm waiting room, with nice carpets, all in multi- coloured blues and dusky pinks. When the nurse calls out your name you tend to jump a little, well, I did. The nurse said don't worry it won't take long, take as long as you like, just get it right I thought to myself. I had what's called a core needle biopsy, this involves the use of a hollow needle pushed into the site of the tumour, and results in minimal disturbance of surrounding tissues and retrieves a solid, intact sample. This is the simplest method for diagnosis, and is commonly used to diagnose tumours located particularly in the breast. Sounds gruesome doesn't it, but it was done under local anaesthetic so it wasn't so bad. Once again my husband waited for me outside, I was only gone about forty five minutes. As I stepped into the daylight he said with a worried look, how did it go. I thought I would spare him the gory details of needles and anaesthetics in my boobs, not to bad I said, trying to appear calm. I told him, my results would be sent to the clinic in the annex at the local Hospital, for a two o'clock appointment the day after tomorrow. As we drove home, I clutched yet another letter to hand in at the clinic. I had such a feeling of impending doom, not just for me, but for the twins as well as my husband. He didn't say too much about the future or what if, well, you know. Remember when I went for my mammogram I had a walk round the shops, my husband asked me if I wanted to"hit the shops", he tries to be trendy at times, but it doesn't work, I like him as he is, a bit, Greenwoods, or Marks and Spencer's. Ten minutes to two, on a wet September day as I said in the beginning, the entrance to this former workhouse is crying out for a coat of paint and some tender loving care. That pale green waiting room with the odd chairs were of little comfort as we sat there just staring at the walls. I'm glad we didn't have long to wait. We were shown into a much brighter consulting room, with just a hint of comfort as we sat in simulated leather chairs. The doctor or, to be more precise , oncologist, in his early fifties with beautifully manicured nails and a shock of silver hair and bright blue eyes, reminded me of Cavana Q.C. He got up from his black real leather chair, shook hands with us and introduced himself, then sat down again as he read my letter of introduction and my notes so far, you could hear a pin drop. My mind was just beginning to wander, and think the windows needed cleaning, and how I would move the desk by the widow, anything to take my mind off the reason we are hear. I looked back at the doctor just as he finished reading my notes, he asked us a few questions about our lifestyle, how long had we been married, did we have any children, questions I'm sure he knew the answers to already. The tone in his voice changed to one of sympathy, but not sycophantic, as he told me that the biopsy results confirmed that both breasts had abnormalities, and that he could confirm that I did have a cancerous tumour in my right breast, and what could be an early stage tumour in my left breast too, he said the biopsy on my left breast was inconclusive and would need further investigation. The worst two words in a doctor's vocabulary, "breast cancer" two words, twelve letters. However it's wrapped up it means the same. It was of little comfort to me as he went on to say it was very rare for a woman to develop tumours in both breasts at the same time, be they malignant or benign. It's nice to be different, not! What the hell do I do now. My first thought was, how long have I got, My second thought was, how long have I got. He stood up and came round to the front of the desk, as he sat on the edge he took hold of my hands for a moment, never have I felt such soft, but at the same time, strong, masculine hands, his bright blue eyes seem to say to me, have faith, believe in yourself. Those few moments as he held my hands were more comforting than any words could ever say. As he went to sit down again I was beginning to feel a little better. He said the first thing to do, was for me to have one or two other tests just be sure the cancer had not invaded other parts of my body through my lymphnodes. He went on to tell me that as the anomaly in my left breast was still tiny he would send me for a scan to try to get a better understanding of what we're dealing with. Now I am getting worried. He said my treatment would start with the C.T. scan tomorrow, then when your ready, we can work out a regime for your treatment and determine how much surgery we need to consider. He went on to say they would also remove the lymphnodes from the breast tissue and under my arms as is usual with this procedure known as a modified radical mastectomy. He said the chemo' should eradicate the anomalies in my left breast, thus making my treatment so much less invasive. He said it was as much down to me, and my attitude, how well I do, and not just the treatment presribed. Not something a girl wants to hear, but what the hell, lets go for it! Having a breast amputated, and make no mistake, that's exactly what it seems like, but in actual fact it's not at all like that, but nevertheless it's quite traumatic and something I couldn't begin to imagine. Yes I know I was swanking about being a biology teacher but I'm not too sure about this branch of biology, it's not something my sixth form girls want to learn about at sixteen, and this is not a medical trip round my boobs either, it's an account of my fight against cancer and the best way to deal with the situation. I tried to cheer myself up by thinking, well at least I would loose some weight because I was quite well endowed, but it didn't work. Truth be known I was frightened to death. He said, four to six months is considered average time for my treatment and recovery, but as I was quite fit for someone of my age, "cheeky thing". he went on to say, I could go back to work about three to four months after my surgery, if I felt up to it. Believe me, work was the last thing on my mind, but we thought that he was just saying that to make me feel better. Why would I go back to work so soon, I can apply for extended leave on full pay. That's something I must look at some time soon, so why bother, my health is more important. He said he couldn't give an exact start date for my treatment, but would expect to see me very soon, certainly in a week or so, he said you'll receive a letter from this department in the next few days. I think this delay was for my sake as much as any bed availability, just to let me and mine get used to the idea and make any domestic arrangements, you know, kids and work stuff.

As we left this doom-ridden satanic looking building I couldn't help myself, I wrapped my arms tight around my husbands neck and sobbed my heart out. Strangely, it was a comfort to feel my husband crying with me. As I said on the way in we had to hop and skip round the puddles, not something I could do on our way out. I just walked through them all and got wet through. I thought sod it! what the hell does a pair of shoes matter. That was the first time I can remember actually feeling sorry for myself. On our way home we had to decide what, and more importantly, when to tell the boys. They didn't know about my biopsy, or indeed anything about my cancer. We decided now was the time to tell the boys, just in case anything went wrong. They're both studying biochemistry at Bristol university, following in their mum's footsteps, that makes me and their dad so proud. They're not due home 'till Christmas, so we decided to tell them this evening rather than wait. I thought it would be best if I 'phoned them, hoping they could tell by my voice that I was felling fine, and do my best to reassure them that everything would turn out for the best and not to worry, and as ever, looked forward to seeing them both at Christmas. We decided to ring after tea in case there were going out, or wanting to study. It's handy they share a flat, I only need to make one call. One call that had the potential to either make them weep, or rise up to the heights of stoicism, and show their mum that they are growing up into fine young men with the potential to go far. I tried to play down the seriousness of my illness, but they're not stupid, they knew the score. As their dad said goodby with a few words of comfort, I heard him say, the best thing you can do for your mum, is to get stuck into your studies and make her proud of you both, as any dad would say. We're not particularly God fearing people but we, like most of our contemporaries believe that something, or someone has a guiding hand in what we do, and what life may hold in store for us. Someone that we thank in times of good fortune, or berate when thing don't always go as we would hope. At this particular time in my life, to handle such devastating news as this, well, I wouldn't wish it on my worst enemy, not that I have any. I just didn't know what to do, or how to prepare myself. Should I prey or would that be to hypercritical. My poor husband was at a loss as to the best way to help me, even after all those reassuring words from the oncologist he watched me sink into deep despair. It's just as well our two sons were away.

After a few days I managed to gather some strength, and with the love and comfort of my husband. I got through this feeling of despair, and found the strength and determination to get on with my treatment and try to beat this 'demon'. The first part of my treatment was a course of chemotherapy over the next couple of weeks. I was told, what I already knew, that most of my hair will fall out, but there was a chance they could save most of my hair loss if I was to wear an "ice cap". But there are varying schools of thought as to weather it could do more harm than good as my chemo' is a slow process. I'm not sure if I did the right thing, but it did save me the indignity of wearing a wig even though by head was numb for ages afterwards. I did wake up some mornings to find more hair on my pillow than I would have liked, and having a shower was always distressing, to see my hair being washed away like that, all I could do was to wear a swimming cap. After speaking to a colleague at school, yes I did tell them, and gathering anecdotal evidence, and lots of reading, and a word with my oncologist. We, meaning my husband and I, decided to go for a double mastectomy. A bit rash you may think. But think again, can you just imagine a year from now. Another lump. Need I say more. The oncologist was very sympathetic and thought I was being a bit radical, but would be willing to perform double mastectomy, so long as I was absolutely sure it is what I wanted. I said we had had a long talk about it, and feel that this is the only way forward so lets get it over with. Being a biology teacher, you would think I would want to know all the in's and out's of my treatment wouldn't you, but I couldn't be bothered to ask to many questions, maybe later. I had enough to think about right now. I suppose it's a bit like the dentist telling you all the gory details of tooth extraction whilst your sat in his chair. All this sounds horrendous I know, but I feel sure this is the best chance of ridding my body of the cancer. Needless to say my operation was a complete success according to the surgeon, and all those involved. I don't think I'll go into any detail at present, but as I said, once you have cancer the chances on ridding the whole body of any cancerous cells completely, are virtually zero. It only needs one microscopic cancerous cell to pass through the lymphnodes or blood supply, to re-ignite the cancer cell growth in another part of the body, the good news is, that my lymphnodes have been removed, cutting off at least one route for any stray cells to wander. Head 'em off at the pass as my husband said. I felt very optimistic about the future, even if my cancer is going to spread, it may not manifest itself for years. I could get run over by a bus, or die of something picked up on holiday long before then. They say there are no atheists on a sinking ship, well, its true. I preyed and preyed to this, God, this, guiding hand, like I have never prayed before, and it seemed to work. A few weeks after the operation, and several more doses of chemotherapy, I felt much better in myself and greatly relieved, and thanked God or whoever for my deliberation. As the sickness wore off and I felt less grumpy, and believe me, Chemo' really dose make you feel very sick, in fact, I thought I was going to die sooner rather than later. But when the sickness did at last subside I began to feel a lot better My oncologist told me, because I acted immediately, had the surgery and chemotherapy, he considered I had the best case scenario for a complete recovery. If I keep up with my physiotherapy and look after myself, he felt sure he could give me a clean bill of health in a week or so, and don't over do things. I had a difficult time just after my surgery, it felt very odd, trying to get used to my new padded bras, and not having any wobble as I used to when I ran for the bus.

Three weeks before Christmas, the familiar white envelope sporting the *N.H.S.* logo, dropped on the mat. I'm glad it was early because my husband hadn't left for work yet. As I took a deep breath, and sat down, like a coward I gave it to my husband to open. The expression on his face said it all. He didn't need to read it to me, I just jumped for joy, till he said, not so fast, you need to have another scan just to be absolutely sure you're clear of the cancer. Then after the results, he wants to see you to have a quick look at you before he can say anything definitive. How soon I shouted, impatiently. Next week for your scan he said, when next week, just a minute, the fourteenth, then the sixteenth to go and see your oncologist, great I shouted, as he past me the letter. The best Christmas present I could wish for was a clean bill of health. Once again, as we sat in that little, pale green waiting room, by now a lot colder and still with the same magazines, colder, because it was only two weeks before Christmas. Why hear, and not the hospital where I had my treatment I don't know, and what's more I don't care. This time, what seemed to be the usual ten minutes felt like a very long time indeed. We were both happy to see our very own silver haired friend as he called us into his office. But this time we both knew instinctively that the news was good, and it was. He said there was no sign of my cancer or any anomalies of any kind, but he still wanted to see me again in a few weeks, then every six months thereafter. As we were leaving, we realised what we had just been told. That little green waiting room lit up as if by the heavens. That doom-ridden satanic workhouse, wasn't so bad after all with it's shiny red bricks and beautiful oak doors. This time I didn't need to hop and skip over and around the puddles, because there was a big orange ball in the sky, this time shining down on little ole' me.

As we drove home everything was right with the world and I felt bloomin' marvellous, I couldn't wait to 'phone the boys to tell them the good news. As I was just about to 'phone them, guess who came through the door, yes you've guessed it, they left university a day or so before the end of term to get home to see their dear old mum, and the look on my face must have conveyed the news far more eloquently than any words ever could. I couldn't help but cry, but they're grown up now, and grown men don't cry, but I think they were crying with joy inside as they gave me the biggest and best-est hug any mum could ask for. We all laughed and had a great Christmas and looked forward to a better new year. I was soon back to my old, well, not so old self and felt much better. In the new year, well more like mid January I went back to my teaching job for two mornings a week. With my new padded bra's and wool sweater I really looked well for woman with virtually no breasts. I was so happy to be getting back into a routine once more. As I was still a bit tender across my chest and was afraid to get on the bus, my husband insisted he take me there and pick me up in the afternoon, just to see how I felt. A supply teacher had been hired to help me, but I needn't have worried, I felt fine, but to my surprise, the local education authority wouldn't let me do more than two mornings a week till after the Easter brake, and only then if I passed a medical just to be sure. They didn't want any come-back in case I fell ill at school. I was soon back in the old routine, and routine just about sums it up. My biology teaching job became easier and far less demanding. I had been doing it for so long I was beginning to get itchy feet and started looking for something more challenging. No! not a hobby, a completely new career, a new direction in life. After my cancer scare and subsequent clean bill of health I felt I could conquer the world, or at least have a damn good try, I wanted to try my hand, and my brain with something completely new. I have had, for a long time, a feeling that I could do more for my fellow man, and woman of course.

A NEW BEGINNING

Before my illness, I was beginning to take an interest in local politics but didn't do much about it, except grumble at the tele' and newspaper articles as most of us do. My husband said I was in danger of becoming a grumpy old woman. Then, one day, in April 1996, close to the local elections to be held in May, a local party candidate knocked on my door. I felt so sorry for him, with the remnants of the latest April shower dripping off his cap, and his umbrella looking worse for wear, I couldn't help but ask him in for a cup of tea. We had a long chat, "Of shoes-and ships-and sealing wax- Of cabbages-and kings-and why the sea is boiling hot-And whether pigs have wings", Oh! and politics of course. Not only at local level, but nationally too, we seemed to hit it off right away, and agreed on many points. As he left, he thanked me for the tea and gave me one of his leaflets and invited me and my husband to join their party as they needed people like me to join them, well! I was very flattered that they thought we were suitable people for their kind of politics, particularly at a local level. I know what your thinking, it's not all that long since my opp', but as I said before I felt great. After the local elections we joined our local branch and began to attend meetings and help to raise funds for the next time round. I just wondered how far I could go with this politicking lark, but my husband was adamant that I take thing slowly at first, but I was beginning to enjoy it so much. My husband was very supportive of me but when it came to attending my local area committees he was not quite as keen as I was. As I got more and more involved, I began to give more time to politics. Then, to my delight, I was invited to join one of the committee's to promote a particular initiative, I jumped at the chance to show them what I could do, and more and more seem to put my husband in second place, but not once did he complain , not even when I put my name forward to see if I could become a councillor for my own ward.

One Friday afternoon I turned into our drive to see a little blue Minicar in the drive. Who's this I thought, I wasn't in the mood for visitors after a particularly hard day, as I squeezed passed the wretched thing, the sound of the horn, made me jump out of my skin and was just about to give the culprit an ear bashing, when my husband pushed the passenger door wide open and said get in. I thought to myself, I know petrol has gone up again but this is one step too far in the economy drive. Where's your car, why have you swapped it for this little buzz box. My car's in the garage, what's wrong with it, nothing he said, I mean our garage, this is yours, mine, yes yours, get in we'll go for a spin round the block. With a million thanks and even more kisses we were off. With a refresher course at our local driving school I was ready to go. A perfect start to my first foray into politics, this time on a more responsible level. The next aid I needed for my new career, was a computer and printer. A real alien concept to me, but my husband was getting to grips with computers at work and was able to get me started, a few lessons from a colleague at school, I was soon surfing the world wide web, sending emails left right and centre. I had to write my own election leaflets, appoint an agent, I could be my own agent, but it looks better if I can say "my agent" will see to this or that, good ah'. I managed to gather round me quite an army of helpers including my husband. Delivering all over the ward, putting up posters, we managed to find 45 poster sites in just the one ward, that's very good. Whilst it was hard work, what with my teaching job, my duties as a school governor, running around like a blue arse fly, it was all very exhilarating. What I didn't expect, was a call from the practise nurse at our local health centre, asking me to call in the surgery the next evening as my doctor wanted a word with me. What ever for I asked, but she said she didn't know. I pressed her as much as I could, but she could only repeat what she had already said. Well, as you can imagine I couldn't wait to 'phone my husband, I was on pins as I waited for my husband to pick up the 'phone. After what felt like an eternity, a heard a man's voice, for a split second I almost blurted out about the doctor's call, but just managed to stop myself. The man knew by my voice who I was, and said my husband had just left the office and would be with you soon. Soon! it would be at least half an hour before he gets home. To my shame I was still quite weepy as I heard his key turn in the lock. Whilst he took his coat off I made a cup of tea and told him about the phone call, and how I tried to find out why she wanted to see me about at such short notice, he was equally perplexed. If she thought it was urgent why can't we go now, or in the morning. All he could do was to comfort me, and said it may be something and nothing, how can it be, I've only just been to the clinic for my check up. Could it be something to do with that, had they discovered something that I should know about. This was just about the longest twenty four hours imaginable. I rang the surgery several times during the evening, but couldn't ask the answer 'phone what it was about. The next morning, didn't feel anything like next morning, I hadn't been to sleep. I was more worried than at anytime during all of my treatment. Remember, the lump, the mammogram, the biopsy, the finding of the tumours, the mastectomy. Self examination is difficult after a mastectomy, I did try, but It was no use, I couldn't feel anything at all. There was no way I could go to work that day, I asked my husband to ring the school and say I was unwell. As luck had it my husband said he could finish work at three o'clock, and was home by half past. As we sat at the kitchen table we, well, I went back to the day we sat in the car watching the rain and wondering whats in store for us, and you know how that turned out. Five o'clock came sooner than one would expect in this situation, but here we are, on our way to hear what the doctor has to say. As we sat waiting for the doctor to see us, I fearing the worst, my husband hoping for the best, both with nothing to say. Then we were suddenly jolted from our day dreams, more like nightmares for me. On this occasion I asked my husband to come in with me. We sat down to hear what the doctor had to say, after all night, and all day, worrying myself sick, the what ifs, am I going to die. All she wanted to say, was that she had obviously noticed that I am standing as ward councillor, and that I should not over do things, and to take it easy.

She went on to ask how I was getting on personally, not party political of course. She said I looked very well, then we had a little chat about diet and life style, how I was coping at work. I think she wanted to take a look at me at close quarters, you know what they say, one picture paints a thousand words. She went to open the door and said good luck with your campaign and good evening, I'm sorry to tell you this, but my husband blew his top. He told her she was just about as insensitive as was possible to be, he went on to tell her that I had been worried sick all night, just so you could "have a word" how could you be so thoughtless. Why didn't the receptionist say it was non urgent. When he came out he said sorry to me, but said the doctor needed telling off. As we walked to the car, the relief was unbelievable, never did I stop to think she just wanted to see me and make sure I wasn't over doing things. Wow! I said to my husband, me thinks a chicken curry and fried rice is in order, good idea he said. Make sure they're double wrapped, we don't want to contaminate the car with curry smells.

After weeks of campaigning, delivering leaflets by the thousand, going to work, gobbled food, postal votes, knocking on doors, all with a spend of only £789 on my campaign, I won a seat on the local council, May 3rd 1997. a day I'll never forget as long as I live. Hope it's a long time. On a council with a majority of my party's members we were in charge. I did pretty good for a girl, re-elected for a second term, quite a good spread of portfolio's and a good grounding in politics as a whole. All this politicking, on the council, my teaching job, school governor, it was damn hard work, but I loved every minute of it. I heard, quite some time after joining the party and winning my seat, that they had in fact been watching my progress with interest, not just my illness, but my work ethic and my determination to help others, and to raise awareness of the terrible toll cancer can have on people's lives and families. One of my proudest moments was in 1998 when I crossed the finishing line of the London marathon. We managed to book a couple of nights in a little hotel down a side street, don't ask me where, they all look the same to me. You may have seen on the tele' the pasta parties the night before the marathon. We called in on one of them for some free pasta as you would, but we left early it was all I bit too hectic for us. So it was back to the hotel, and an early night, no, not that kind of an early night. Sunday 6am. I usually get up early, but 6am is pushing it a bit. I don't wish to relive the aches and pains of the marathon itself, just to say it felt like the greatest achievement ever, well having twins takes some beating, but you know what I mean. I crossed the finishing line in five hours, one minute. My husband said, couldn't you have made in under five hours. I could have hit him, but I did squirt water down his neck as we hugged each other. We didn't do much sightseeing I was too knackered, but we had a very nice evening walk round, the weather was beautiful as we sat on the embankment across from the Palace of Westminster. I didn't say anything to my husband, but I couldn't help having the feeling that, as somebody once said, "I'll be back". In a way I was back, but not quite as I imagined. I did yet another marathon in 1999 This time I was a year older but still managed to come in at just over five hours, by eight minutes, not bad for someone not that long after cancer, what was I thinking of . Well, I did raise lots of money for cancer charities in my home town and particularly that "grey" hospital annex, with the cold green waiting room. I felt so good, I wanted to say thank you by way of a donation to those who were so good to me, in my darkest hours. I'm going to skip on a bit, to autumn 2000, don't want to bore you do I. I did very well as a councillor at a local level, well, I like to think so, and soon began to set my sites a little higher, well, a lot higher, no less than the Palace of Westminster. I really fancied the idea of being a member of Parliament. The next General Election was looming on the horizon, well, not quite the horizon, it could be as far away as 2002. But rumour has it, it could be as soon as May 2001 so there was no time to dither. The constituency where we had lived all our married lives was crying out for a new face on the political scene. I had a word with my husband, and of course the twins, and with one voice said, go for it mum, it's better to have tried and lost, than never tried at all, sounds familiar! As I wasn't sure what to do about it. I approached the party chairman, who in turn spoke to one or two members to canvass opinions, and very soon was happy to give me the green light, even though I don't need his permission, it helps if you can get a good reference. With my C V and a cheque for the £100 fee, a letter was dispatched post haste to London. After a week or so, word came back in the form of a posh letter asking me to attend an interview at their London office, and would I e-mail them to confirm receipt of their letter. Try and stop me I thought, and with the blessing of my husband, and the twins, a new posh frock, I caught the 8.23. train to London.

It was five years to the day since that fateful day in

the shower, and the memories of that little green waiting room. I wonder if it's been painted yet. As I watched the world flash by, I must have dosed off for a few minutes, I wish I hadn't. I had a very odd dream, I could hear people sniggering at me behind my back for thinking I could be a Member of Parliament when I was still wet behind the ears. I was glad when the train jolted me from my nap, and thought whoever is up there trying to spoil my day, stop it now! At that point I went to the buffet car for a coffee and a big sticky bun, sticky buns work wonders for me when I'm down, well, that's my excuse anyway! I was soon back in my seat, thinking about the interview and what to expect, but I needn't have worried at all. Even though it was a very demanding interview and a test of thinking on your feet, answering quick fire questions from left right and centre, writing briefing notes on topical political issues, all without notice, phew! But I felt I did pretty good for a girl. I don't want to get to big headed, but with my background in education, and a good record as a school Governor, and successful councillor, well, I thought, can't go wrong, and felt quite good about myself as I made the long journey home in a sunny train on a sunny autumn evening, a journey that was to become a lot more familiar in the coming months and years.

The postman calls early in our district, this time he knocked on the door and said he couldn't push the large brown envelope through the letter box because as it was a bit to bulky, I couldn't help but grab the letter from his grasp, then I had to apologise for my action, and thanked him for his consideration. The letter, more like a parcel really, I emptied the contents onto the kitchen table with pages spilling out onto the floor, but no letter about my interview, I began to think I had fluffed it, but there in the folds, was that most important piece of paper, the one that begins, 'we are pleased to inform' you wow! I was accepted on the candidates list for the next general election, it was the best news since, well, you know when, it felt like one of those times to thank Him or Her upstairs whoever He, or She, may be, but I think this time it must be She. I hugged my husband so tight his neck creaked. Being accepted as a candidate is only the beginning. It means I have an opportunity to go before a selection committee, but not necessarily in my own constituency, and hear-in lies the rub. If I was to be offered an interview in a constituency that was a distance away, what should I do. It would mean if I was lucky enough to win the nomination, and subsequently the seat, then my husband would have to give up his job, and resettle in that constituency just to please me and my ego. Something we hadn't really given much thought to. Every evening, when I get in from work I boot up my computer, sounds good ah! and make a cup of tea. As I looked at my in box I literally jumped for joy. Lo and behold! an email from our own constituency chairman. I couldn't believe it, the email said the constituency had been given permission by head office to advertise the candidate vacancy, and invited me to make formal application, and saying I would have a place on their list with a strong recommendation that I get my letter and C.V. to them as soon as I can. That was the best possible outcome to my dilemma, although I didn't tell my husband. The email went on to explain what I had to do next, such as compiling a new C.V. more suitable for the constituency. I wasn't to sure what he meant, but dare not ask, I didn't want to look stupid. My constituency wouldn't treat me any differently than any other applicant, nor would I want them to, well, maybe a little bit. As with any constituency, I would have to battle it out in front a selection committee in a few weeks time, hope to be put forward for the next round, and then do my stuff in front of the whole constituency membership. Maybe as many as two hundred people. Fortunately being offered the opportunity to go before my own constituency selection committee was a big help, but I've not been this nervous since, well, you know. Yes another new frock, a nice warm plum red with grey trim and grey show buttons down the front, a zipper at the back, black shoes, black tights, and a big smile. As my husband drove into the car park, on a cool Saturday morning in late September, he found a spot near the door, we dashed in from the cold wind and were met be the chairman and one or two members. About a week before the selection committee is due to meet, the candidates names are put in a hat, and lots are drawn to arrange the order of appearing before the first selection committee to do our stuff. Each candidate was sent an email with the time of their expected presentation. It's up to them how, and what time they get here. I'm not sure, but I always feel that any selection committee will be getting a little board by the time the last person make his or her pitch. There were six other candidates applying for the chance to contest the seat at the next election, be it in May next year, or 2002. Because I didn't want to be thought of as having an unfair advantage, I kept myself to myself as much as I could, as we waited in turn to make our pitch for the chance to go through to the afternoon session. Being a teacher gave me a bit of an advantage, or so I thought. I'm glad my husband had his wits about him, he dose sometimes. He said to be careful not to talk down to them as if I were teaching a bunch of sixth form girls, good thinking batman! Nervous as I was, I did well enough to go through to the afternoon session with two other candidates, both men. During the lunch break we had time to canvass opinions, talk to the other candidates to try to see if we could see any weaknesses. I would have loved to have gone out for some fresh air, even though it was bitter cold. Instead we "networked" the room as much as we could, to try to gain as much support as possible, the one thing I didn't do, was to by anyone a drink, that would be a very bad idea indeed, it could look like bribery. I felt very confident in myself, and I'm sure my hubby did too. The currant incumbent has been our M.P. for three terms of office. It's about time we had a change. Once again I felt I did well enough to take it, and I was right. I'm not supposed to know, but I was told I was very good and created a the right impression. but I only just edged it to go through to the next round in the afternoon. I found that a bit worrying. This time we drew lots to see who would speak first. As there were only three of us going through to the next round, it didn't matter that I was second, sounds like a game show, but believe me, it's no game. We all shook hands with each other and waited to be called. I had no idea how the first man did as I went in to do my stuff all over again. I had to stick to the same sort of presentation as this morning, but vary it just a little so as not to repeat myself to often to those that were part of the morning session. As we waited for the last one to do his presentaion, then the ballot by the members, I allowed myself an indulgence by way of a large gin and tonic with ice and lemon, and a pint of best bitter for his lordship. After about twenty minutes the chairman came out of the concert room and called the three of us into the little office near the stairs. As he reached out to shake the hands of my two opponents I realised my greatest wish had come true. I'd won! me, the P.P.C Prospective Parliamentary Candidate. Wow! With hand shakes all round, the meeting finished about four o'clock so we had a very pleasant hour or so, surrounded by many new friends, and I hope lots of helpers when the election is called. The afternoon was so pleasant we had to get a taxi home. It was Autumn 2000. and no use trying to make any plans at this time as we hadn't any idea when the general election would be called. Monday morning, I came down to earth with a bang, and got back to work, I put the election to the back of my mind, and life felt good, very good.

PATIENCE GIRL, PATIENCE

I said I put the election to the back of my mind, but it couldn't come soon enough for me, oh! don't get me wrong, I love my work, all the girls, and my colleagues at the sixth form college, it has been my world, my work, my life, for nearly twenty years, but, as I have said it's time to move on. Christmas came and went, and with the new year came a new hope of better things to come, like a general election, and me as the new member of parliament for instance. The weather was cold and miserable, but spring was just round the corner. Eventually there were rumours that the general election could be held sometime in May now that the lorry drivers were back to normal, and people were beginning to forget the past and look forward to the spring and summer. It's a good job I didn't dither! I was like a seventeen year old school girl, looking forward to university, but ten, no, twenty times more exited. My poor husband didn't know what hit him, with organising here, organising there, organising every bloomin" where, I was so exited, and I am sure, my long suffering husband was too, but he didn't show it, I did wonder if he was bit jealous of me, but not for long, I soon came to realise he only wanted what would make me happy. I didn't know at the time, but nor did anyone else for that matter what was in store, no one could imagine just how devastating it would turn out to be, and how it would effect all our lives over the next weeks and months, the way we travelled and where we could take our holidays. You're wondering what it is aren't you. I remember the day quite clearly, it was 19 of February, our wedding anniversary. The newsreader reported a case of foot and mouth decease at an abattoir in Essex. It didn't click at first, but if the rumour were true about the election, being held in May this year, 2001, an outbreak of this potentially devastating bovine decease could make an election, a non starter. It's not that long since the lorry driver's strike! Now this!. We just had to wait and see how things developed. It could be postponed for months, or even till next year. Elections are usually held in spring because the weather is more conducive to outdoor meetings, canvassing, and delivering leaflets. The Prime Minister had no alternative but to put off the calling of the election for a few weeks, if indeed there was going to be an election. We would have to see how the outbreak was effecting travel, and movement of people and cattle in rural areas. My heart sank as we watched the news, my husband was just as concerned as I was. But as I had hoped and wished for, the Prime Minister did eventually announce in early May that he would seek a second term in office, and ask the Queen's permission to dissolve parliament and call a general election. The poll to take place on June 7th. It felt like a starting pistol had gone off ! I have never been so exited, apprehensive, afraid, all at the same time, but raring to go nevertheless. I couldn't wait to get going, but the T.V. pictures did nothing to help the situation, with burning cattle, and talk of foot and mouth every where we went. Trying to canvass was hopeless, no one wanted to talk about anything but the damn foot and mouth outbreak, and what about their holidays, as if I knew! It made campaigning very difficult for us, after all my planning for the election I must admit, this foot and mouth business was a damn nuisance. But I must remember, if I want to become a Member of Parliament I must be more pragmatic and understanding, so I'd better keep my mouth shut and my thoughts to myself from now on. I'll just have to be patient like everyone else. As I was already in place as the constituency candidate, all I, or rather the constituency chairman had to do was to register me as the Party's candidate, and find me an agent. Someone suggested my husband, but I thought, he won't want to do it, but to my delight and surprise he said he would love to do it. He said he had been reading up on the subject, and felt quite confident about it. With the £500 deposit taken to the town hall, and all the formalities of my nomination papers properly presented, with all the officers appointed, and a fighting fund in a separate bank account, we were off! Canvasing and leafleting, putting up boards, with my name on them. I really felt we were getting somewhere. I spent all my spare time going to public meetings and canvassing, I was out every minute of every hour I could, speaking to as many people as would listen to me. I found it quite easy to speak in front of a room full of people because of my teaching experience, but I had to be careful and not talk down to people, as my husband kept reminding me.

The reception on the doorsteps, once we could get past the inevitable comments about the foot and mouth fiasco, were very encouraging, and made it feel as though I was beginning to make some headway, could it be because

I'm a woman, I would be the first woman to represent this constituency as far as anyone can remember. Men say woman can talk! it's a good job too. I didn't know at the time just how close it was going to be. Of course there are always a few small single issue parties who just want to protest about something or other and pinch a few votes off the mainstream parties, but it's a free country, we looked upon it as an occupational hazard, and just got on with it. We all like to think our campaign is the best and most open, honest , and well run campaign we can muster, well I think we did a pretty good job of it. After lots of hard work on the hustings, door to door canvasing, photo' calls and political pieces for the local newspaper, what wouldn't I give for five minutes on the television. All the foot slogging, not only by me, but my poor long suffering husband. Oh! by the way, I didn't tell you what he did for a living, well, I'm not entirely sure, but a you know, he's computer savvy, and is something to do with corporate entertaining or organising, something like that for one of the big chemical companies. For once the foot and mouth outbreak worked in our favour. As I said my husband was in corporate entertaining, but what I didn't know was that most of the products were for the farming industry, and as such rep's were banned from farms and the company site. That being the case was great! he managed to get a couple of weeks off so he could help out and generally make himself useful. Talk about silver linings. He was very good at getting the best out of people, not least, our deliverers, and not forgetting to thank them from time to time. With dozens of helpers delivering leaflets galore, something like a hundred and twenty boards in people's gardens, numerous paper posters in windows all over the place, the sun shining, the birds singing, it was great!. Then it hit me, I saw a private ambulance from the local hospice sporting the cancer research logo. It's a good job I was on my own. The site of that ambulance was an untimely reminder of what was, and what could be. As I felt the warm sun on my face, I thought to myself, enjoy it while you can, and carried on down the road with a hand full of leaflets and a heart full of hope. After three weeks of frantic walking, running, talking, listening, food on the hoof. Plus my job as a teacher and my council work, and all the work done since I was selected. Delivering leaflets during the winter and spring, in all weathers, at nights and weekends by us all, by all, I include my husband, and all the helpers in varying degrees, but they are all my friends to the same degree. A group of us met at the club on Wednesday evening to plan for the final hurdle, polling day, and what a day! I was running around like a don't know what, all my helpers played a blinder, I felt so confident. By five o'clock we thought we would get off home and get washed and changed for the evening push. But of course we hadn't prepared anything for tea so we collected a fish supper on our way home. That fish supper was the best meal we'd had in ages. After a quick shower, and my new frock, yes I treated my self, again, if I won, I would look the part, and if I lost, I would still look the part.

Ten o'clock. The polls closed, and off we were to the town hall, well, not quite, we nipped home for an hour, we felt we had earned a breather. Just to sit down and have a brew and go over the days events would be nice. But I was far too exited to sit at home no matter how tired we were. I dragged my poor husband out of his chair and off we went to the town hall. As we arrived we were showered with hugs and kisses all round. I hadn't won yet, what's it going to be like if, sorry when I do win. The count is always nerve wracking at the best of times, even if you feel sure you've won. But this election was not over by a long way. The first count was far to close for my opponent to be happy with, and asked for a recount, as is his right, I was happy with the count, it had me down as the winner. The second count was no more decisive than the first, but if my nearest rival wants another count, it's up to him and the retuning officer. But my opponent leaned over the table to shake me warmly by the throat. He was so deflated having lost by such a narrow margin, At last! nearly two o'clock, the result! what with the recount, frayed nerves, and everyone totally knackered, the returning officer called all the candidates to the council chamber, we knew the result already, but listened in total silence just so we could savour it all over again. As the results were read out, and a win for me, it was hugs and kisses all round, just a big hug from the boys, you know, I was close to tears as I could see my husband and our sons looking so proud of their dear old mum, as we went back into the ballroom, I climbed the steps up to the stage and cast my mind back to that fateful day just over six years ago when I thought my life had come to a premature end. My first public speech, as an M.P. Well, not quite yet, I've got to take the oath down in Westminster, but how exiting anyway. I thanked the other candidates for a good clean campaign and wish them well in their future endeavours. I then thanked the returning officer and all the town hall staff and of course, the police.

BLOOMIN' MARVELLOUS!

Only a few dozen votes between me and the incumbent member of Parliament, but never the less we won. I must say I have never known such an intense, and at the same time exhilarating few weeks, an experience never to be forgotten for the rest of my life. I feel so happy, I should resurrect Dr Johnson to invent some new words to express how I am feeling right now, because I can only come up with, is 'bloomin marvellous'. After a few late drinks at our local club, 'don't tell anyone', to celebrate what was a stunning victory not just for me, but for all of us. We went home that night and had the best nights sleep in weeks. The next morning, Friday, was as sunny as anyone could wish for, as my husband went to collect his car from the club, I spent hours on the 'phone thanking, and being thanked by my many friends old, and new, life was great. After lunch I decided to go into school, I couldn't wait to see their reaction, but at the same time didn't want to be seen as swanking, but I couldn't resist. All the girls, and my colleagues with whom I had worked for quite some years, and must have bored the pants off at times during the run up to the election were delighted, and at the same time, came to realise that I would soon be leaving them. My own class in particular said they would be sorry to see me go, but nevertheless they were happy for me and wished me every success, as they presented me with a beautiful black real leather briefcase with my initials in gold leaf. They were a little embarrassed to say, but they hadn't put M.P. on, just in case I didn't win, but would have it ready for me before my departure next Friday. I was contractually obliged to stay on at least till Friday. But I felt confident that I would win the seat and hand in my resignation when I won, and resign my post, to that end I had contingency plans in place, with all the appropriate letters ready to hand in, all the girls course work sorted, and my desk cleared ready for the new teacher, I couldn't wait for Friday. It was only five days, but it was more like five weeks. After a drink with my colleagues and a tearful goodbye to all my girls as they gave me my more official looking briefcase, with M.P. in gold under my initials. I almost ran to my car, threw all my stuff in the boot, and off. I'm a bit ashamed to say, I never missed that school one iota. My husband and I had a most joyous weekend since our wedding, with the birth of the twins coming a very close second. I was going to take my seat in the House of Commons no less, just think, the Palace of Westminster. A dream come true. We decided to make a weekend of it. This time we decided to splash out a bit, and booked two nights at a very nice hotel near Victoria station. So, here we are five years and ten months down the line. With my cancer well and truly out of the picture. We would soon approach the doors to the Palace of Westminster, instead of those big oak doors of that red brick former workhouse on that drizzly day in September 1995. We caught an early train to London on a beautiful sunny, Saturday morning, as I said earlier I was like a teenager going to university, but much more so. The prospect of actually being a Member of Parliament fills me with joy, excitement, trepidation, and an overwhelming feeling of mission accomplished. After we were shown to our room in what turned out to be a lovely quiet hotel, we couldn't wait to have a look round. We were like two kids on our first outing. Our first port of call was the Palace of Westminster. We just had a walk round the outside, we didn't go in, we thought we would save it for Monday. The theatre on Saturday evening, was a real treat, my husband managed to get tickets for the mouse trap at St Martin's, followed by a first class dinner at our hotel and an early night, if you know what I mean. Sunday was lovely and sunny, we were given details by the party's accommodations office of a flat just over Westminster bridge, with a recommendation to take it even if it's not exactly what we are looking for as accommodation so close to the "House", as we M.P.s call it, is difficult to find.. We were given a 'phone number to call as soon as we arrived in London, this we did, and arranged to see it on Sunday afternoon. Here we are on our way to view this second Palace of Westminster, not. It wasn't too bad, more like a bed-sit. The problem was, we couldn't move in till Monday week as it was being fitted with a new bathroom and kitchen. I don't mind waiting if it means a new bathroom and kitchen. Believe it or not the rent of £950 per calendar month, by all accounts, is quite reasonable, I can easily pay from my parliamentary allowances anyway. We had a very nice day by the river, just soaking up the atmosphere, with all the sounds from the river, the foreign languages, the colourful costumes, it was great. A ride on the millennium wheel made a good day, great! I couldn't wait to see the twins tomorrow morning. They said they were so proud of their mum they wouldn't' miss my, "Triumphalist" march into Westminster for the world. We had arranged to met them just across the road, on Bridge Street about nine thirty. I was really sorry my mother couldn't make the journey with us. But we'll take plenty of photographs for her to see later. It had been arranged for me to take the oath along with some other late comers or new intakes as we are called. The ceremony is to take place immediately after prayers on Monday morning, about ten thirty. So another new frock, a nice light plain cream cotton pinafore dress, with nearly black tights and black patent leather Cuban heeled shoes, and a tiny black clutch bag. I felt great. After what seemed to be endless security checks, we were taken to the house administration office for our passes, after the form filling we were shown to our leader's office to meet all my new colleagues. The first person to meet me, was to be my mentor for the next few months, and I hope my friend for many years to come, he was so nice. He introduced me to a number of new Members, then someone, sorry I can't remember any names I was to caught up in the moment, to take much notice who was whom, explained the proceedings to me with the aid of a little floor plan of the chamber, what to do, where to stand, and to look solemn as the ceremony is taken very seriously indeed. I was one of about a dozen new members, from all parties, waiting to take the oath. It was handy really, I managed to manoeuvre myself over to the right of the queue so I could see how it was done, till one of those men in wigs at the top table gestured me to get in line. I had to approach the table of the house, bow slightly to the speaker, take the bible in my right hand, It was at this point that I suddenly thought, is it a him or a her up there......... the speaker coughed a little and brought me down to earth. I gave the speaker a nervous smile as I read the card. " I, Jacqueline James, swear by Almighty God that I will be faithful and bear true allegiance to Her majesty Queen Elizabeth, her heirs and successors, according to law. So help me God. I then had to sign my name and write the name of my constituency in the "Test roll" why roll? I wondered that too. It seems that many years ago it really was a roll, a roll of parchment, nowadays it's a rather grandiose parchment book with blotting paper between the leaves, then I had to step back, bow to the speaker, well not the speaker personally, but the office of the speaker. Then, under my breath, let out one great yelp of delight. I felt so, "phantasmagorical" as I stood there, with my husband, and two sons looking down from the visitors gallery, feeling just as proud as I am at this moment. We were given a whistle stop tour of the many corridors with miles of cabinets groaning under the weight of books, the pictures, photographs, plaques, and bust's of famous people, and not so famous people. The area that impressed us most was the Churchill arch, this arch was built at the behest of Winston Churchill from the stone retrieved after a bomb hit the Commons during the war. We looked at statues of Winston Churchill, Lloyd George. and others, it was a lot to take in at this time. After our tour, a group of us were treated to afternoon tea by the leader of the party on the terrace, you know, the part by the river you can see on the tele'. On our way out I couldn't help myself, I had to take another quick look at where it all happens, the chamber of the house. Well I was a bit naive and soon realised that a lot of work is done in committee rooms as well as the bars and tea rooms, and not just in the chamber of the house. Before we left, the chief whip introduced himself and asked me when I would take my seat proper as it were. I explained I had to wait 'till the flat was ready, and square things at home so it would be next week. He said not to worry as I was not the only one struggling to find half decent accommodation. In the meantime, I'll sort out an office for when you. Come and see me when you're ready to start work. I'm ready now I thought. We said goodbye to the boys as they went to catch the train to Bristol, we set off to walk to our hotel with a spring in our step and a feeling of nothing short of euphoria. As we went into the hotel to collect our cases, as we couldn't take them any where near the commons, the desk clerk asked us if we wanted any lunch, I said no thank you, we've just dined in the house of Commons. How good is that! We had a drink and a much needed sit down in the hotel bar, I couldn't help but close my eyes for but a moment. As I came out of the land of nod, I realised just what I have achieved, sorry, we. There was no way I could have done this without the love support of my husband, and though they don't know it, the love and support on our two wonderful boys. It must be force of habit, but we usually watch the news whenever we can, as we watched the tele' in the hotel lounge, to my delight I saw a glimpse of myself in the central lobby. I called my husband to look at the tele but as he looked up from his paper I had gone, never mind he said, they'll be plenty more times, we must get a move on if we want to catch the three thirty train. The journey home was in some ways a bit frustrating, I just wanted to get stuck in and show them what I can do, but I'll have to wait a while. As I said we were in a state of limbo as we couldn't do anything 'till next week. So we had a few days in the sun. We found a package deal on the net, three days in Gibraltar, glorious, we did nothing but suckle on the sweet taste of success. When we got home the letters were still piling up behind the door, some to congratulate me, and some from people who thought I was still a councillor.

LET BATTLE COMMENCE

I was never a fan of Sunday opening, but this was one occasion we made the most of it. We went to our local super market to stock up on as much food as was reasonable to store, such as tea and coffee dried milk for emergencies, and some tinned beans and soups and other bits and bobs. Then to the electrical shop for a tele' and video recorder, and treated myself to a brand new lap top computer, bedding and cutlery and crockery, and , and, well you know the sort of things I would need. We even bought some emulsion paint, not green! My husband took Monday Tuesday and Wednesday off work, so we could drive down early on Sunday morning with all our purchases and help me to tidy the place up and buy some bits and pieces as and when we needed them, such as kitchen stuff, and a clock-radio we forgot. Over the next few weeks I would make it a real home from home as I added the feminine touch. After three hectic days during which time I nipped over to the Commons from time to time sort out my office, no time for sight seeing, we soon had the flat looking like a little palace, and well and truly Christened, if you know what I mean. I kissed my husband goodbye on Wednesday morning as he went see if his car was still in one piece and where he left it. It's a good job our landlord found somewhere to park, otherwise it would have cost a fortune. It's strange, but you always find out when it's to late, we could have parked in the underground car park at Westminster for free. As I watched him go out of sight I could just see his arm raise up to wave me good by. Then, I turned on one heel, and one toe, and almost marched into the commons and said to myself, "let battle commence" I was allocated an office, well, I say an office, more a half share in a broom cupboard, at the end of quite a long corridor, and no it wasn't painted green, but beautifully oak panelled, yes even in this out of the way corner. It smelt of what I can only describe as, being reminiscent of that smell you get when first entering a church after it has been left closed for a few days, then as you walk in from the fresh air it gives up that smell of pine, polish, books, papers, and years of comings and goings, but in my case, it smelt as you would expect, of the Mother of Parliament, the Establishment this England, Woe! stop right there, just get on with it. I very quickly got into a routine. My office companion was obviously of the same party, she too was part of the new intake as we're called, but she didn't have a shiny black real leather briefcase like mine, hum! We got on famously and not only shared our work related concerns, we quite often went for a snack or a drink together in one of the many bars, and the occasional meal out in the evenings. One day, when we knew things would be a bit quiet. We decided to hit the shops as my husband once said, remember. We caught a taxi to Knightsbridge. We asked the driver how far is it to Hyde park corner, from Harrods as we might have a walk and take a look. We had dared each other to stand up and speak. When he said it was about a mile, that's a mile there and a mile back, we thought better of it. We've more important things to do, like shopping. We had lovely afternoon, looking round the shops with afternoon tea at Harrods, and of course a Harrods carrier bag to swank with. It was a real eye-opener for both of us, just how people can afford to shop in Knightsbridge. Soon after the summer recess, during which time we had a few days with my mother and took her on a number of trips around the lakes, with some really lovely lunches by this lake, or that water. After the recess, probably because of my unenviable knowledge of the devastating effect of cancer and other equally serious illnesses can have, not just for the individual, but family and friends as a whole. I was offered the port folio of junior spokesperson on health. Wow! I thought this was blooming marvellous, but I soon realised that as we were only a small party, most of us got some kind of portfolio, a bit disappointing really, plus the fact that it was unpaid, but you never know, it might stand me in good stead in the future. The months just flew by and soon the nights began to "draw in" and a feeling of work, being, well, work. Don't get me wrong I still love every minute of it, but there are times when I miss my husband dearly, after being married for twenty odd years then suddenly being parted like this, well it was quite a wrench, something we hadn't given much thought to before my election. It began to affect but both of us, as we only saw each other at weekends, even then I had my constituency work to do, and my surgeries, or a meeting to attend, but I did volunteer. I missed him most when I get back to my bedsit, on those cold winter nights after working late, and yes, I do miss our love making. Before my run in with cancer, our sex life was, er, shall we say robust. But obviously during my time battling with my illness as well as my femininity, my poor husband,........... I'll say no more, I'm sure you understand. Since my mastectomy, even though it took some time to get used to the idea of padded bra's and not looking as good in the boob area, as I used to, we had a pretty good time. As I boarded the 7 15am. train one Monday morning, and looked back on the weekend we had just spent together, and realised it was all too brief, I felt like getting off the train and making some excuse to go home to him, but I didn't. There wasn't much point, he would have gone to work by now anyway. A few days later I was telling my friend about how I was feeling, and the fact that I was missing my husband, then, eureka! out of the blue came the obvious answer. My husband could be my researcher. I was sharing a researcher with a busy male front bencher and was always waiting for briefing notes and stuff. I wasn't too sure at first, having my husband working for me, but my friend soon put me right and told me not to be so silly. After all, we'd been married for over twenty years, and should have nothing but a good working relationship to look forward to. I had to agree with her, and couldn't wait to get home that Thursday night. All the way home, the train tracks were chattering to me, wouldn't it be good, wouldn't it be good, wouldn't it be good. I couldn't wait to get home to tell him. I Didn't 'phone him because I wanted to see his face, as I told him, because I wasn't sure how he would react. I almost tripped as I ran up the drive, as I reached the front door, it opened as if by magic, He said he saw me walking with a run, and thought I needed the loo'. When I told him the reason for my excitement, miracle of miracles he was thinking the same thing, how good is that! We both had Friday off, and had a nice ride out in the country, a lovely meal at your typical olde worlde inn, a nice run home just in time to beat the traffic. Smashing.

Unfortunately my husband had to work on Saturday morning, for a few hours, but as I had a surgery from ten o'clock till eleven thirty, at the local library, I said we could meet at the club when he finished work I knew it was a big mistake the minute I walked in. The constituency chairman came straight over to me, Hello he said, nice to see you looking so well, what do you want I said under my breath. I'm looking for a favour, I rang you this morning, but your answer phone was on, so I left a message for you to give me a call as soon as you can. Then I went round to the library, but could see you talking to someone. Well, here I am, what can I do for you. We want someone to read out the quiz questions at your old school to help to raise much needed funds for the school music classes. Shit! I thought, I knew coming in hear would be a mistake. Searching frantically for an excuse not to go, with nothing coming to mind, I had no alternative but to say we would be delighted. Yes we, if my husband is going to be in my employ, he'd better get used to it, only kidding. We didn't have much choice really. When I told my husband he wasn't too bothered. He said we could call for curry supper on our way home, that should eased the pain a bit. I was happy to see some of my old friends, and have a chat about how I was doing now that I had joined the toffs in Westminster. Of course non of my girls were there, they had all moved on to pastures new. As we said good night and wished them well, I couldn't get away fast enough, and wondered how I stuck it for all those years. When we got home we both said more or less at the same time how things have changed over the last few years. The curry was smashing by the way. On my way back to London, the train tracks were chattering back to me, it's going to be good, it's going to be good, it's going to be good, and it was good. My first port of call was the whips office, I asked as many questions as I could think of about employing my husband as my researcher, what are the pit falls, had any spouses, or siblings found it difficult working together, and did he think it was something we could consider. He said there were no problems at all, it's entirely up to you, if you felt happy, so be it, and they would do all they could to help. He said I should go and see the House Administration office finance people, and ask for there advice regarding employers insurance and tax issues like his P45.and stuff like that. Great I thought. After a busy afternoon everything was in place, it was as simple as that. I spoke to my husband on the 'phone that same evening to tell him the good news, he in turn, said he would hand in his notice to his employer, tomorrow. Whilst we were speaking, I heard the microwave ping in the background, so much for being an new man I thought.

In just three weeks we were travelling together down to Westminster to start a new chapter in our lives together. This was to be the best time of all, but he did say he had just one regret. I couldn't for the life on me think what he could mean, having to leave his precious car at home, honestly, some men would rather loose an arm than their car. There, there, I said you can play with it at weekends, so long as you wash my car at the same time. Every time we went home he had to look in the garage to make sure it was still there. I'll just give it a quick rinse, bless him. Things were great for both of us, my husband quickly got into the routine, finding the answers to my myriad of questions, writing up briefing notes and just being there for me. Christmas came and went in a flash. We don't do Christmas any more, not after the way God treated us. But for the sake of my constituents I attend the Christmas morning service with my husband. I sent the usual cards and best wishes as was expected of me, but that was about as much as I could bring myself to do. We managed to get some time in a warm climate for a few days in the new year, we went to Gibraltar again and stayed at the same beautiful hotel over looking the straits. When we got back into the routine and spring started to spring, we had a great time together. We began to settle into a really good working relationship. London is a great place to dine out and see all the best west end shows and art galleries, and all the sights and sounds of such a great city. 2003 was much the same, the same wonderful life in this seething metropolis and because we both worked so well together we could arrange some time off to enjoy the trappings of being a Parliamentarian. As the local elections were due I made conscious effort to help out as much as I could, and we did indeed win another seat on the council. I went back to Westminster, with the feeling that my input had helped the local party to gain that seat. We were privileged to go on several "fact finding trips" they were still work but a most enjoyable part of work. I took the task of helping my fellow man and woman very seriously. Remember swanking all those years ago? We went home every weekend and set to with my surgery on Saturday mornings and would like to think, well, I'm sure, helped most of those that asked for advice, or at least put them in touch with the right people. Another Christmas and new year came and went just as quick as the last one. Don't ask why, I don't really know, but we didn't go away in the new year, we most have been having too much of a good time in London. We were soon well into 2004, isn't it funny how January and February, seem to just fly by. Our routine was never really at routine in the strict sense of the word, more a pattern. The local elections came and went, but not so lucky this time, ah well, may be next time.

STRAWS AND CLUTCHING

In the summer of 2004 I began to get a twinge of back ache, nothing much to worry about I told myself, but deep down I feared the worst. I couldn't bring myself to tell my husband about it, well, it was early days, it could just be the standing about, and my half a broom cupboard of an office, and the rather cramped seating arrangements in the commons chamber. But as time past, I began to feel slightly worse with each passing week, until one morning as I took a shower, I felt a small lump in the middle of my chest, not just a lump, but a bombshell. I feared the worst, and that dreaded word, "metastasis". The spread of cancer from it's original sight usually through the lymphnodes and under the arms, but in my case it turned out to have spread through my blood stream. I couldn't tell my husband what I had found, it would be so cruel to saddle him with such a burden as this, just when we were enjoying life so much. I thought if I could arrange some kind of treatment, under cover of my six monthly checkups without my husband knowing, I might just gain a little time and maybe with chemotherapy things would improve things for me, and maybe, just maybe, he wouldn't realise just how serious it is. But straws and clutching spring to mind. When I told my husband, he was devastated, and shook his fists to the heavens with all the, "a r r r r r gh " he could muster. We just held each other tight, then held each other even tighter. We took some time out for a number of scans, blood tests, prods and pokes. Then the dreaded letter with the familiar *N H S* logo in the top right hand corner. With hands trembling I opened it, but there was no way I could read it, and gave it to my husband, in his soft tones, he said it was a appointment for me at the local clinic to see my oncologist. Neither of us spoke for a minute. Then, rather stupidly I got to make a cup of tea, as if that would cure me, I threw the pot to the floor and just sobbed my heart out. My poor husband was lost for words as he cleaned up the mess. Almost nine years to the day we went again to that, once "grey" hospital. Not any more, the entrance was a lot better, no more grey, all nice and new. The doors had been French polished, and the big brass knob had been polished nice and bright, and the floor relayed. If my visit wasn't filled with so much trepidation, I'm sure I would have been more able to appreciate more what some of my fund raising had done for the hospital, but I was beginning to feel so wretched by now. That little green waiting room was a lot more modern with soft furnishings and modern lighting, but with only a slight change in colour. I'm not sure what it's called but it was a very nice shade of, yes you've guest it, green. I'm telling you this because I am trying to take my mind off what I am sure will be a disastrous day. My husband tries to put on a brave face, but it was so hard for him to be positive, I felt I had to tell him what I thought the oncologist would be telling us. It would be cruel just to let him hear it cold, as it were. I thought it might just prepare him for what I felt sure would be very bad news indeed. Even if one is mentally prepared for bad news, believe me, nothing could ever prepare us for what was to be the blackest day in my life. We were surprised to see our silver haired friend at this, not so posh hospital, but I didn't care anymore. As we sat down I prepared myself for the worst. He said, the small lump in the middle of my chest was just a small cyst, and nothing to worry about, as it would be easy to remove, thank God I thought. Now the not so good news. He said the back pain, and the pain in my legs that had crept up on me over the past few months or so wasn't just back pain, it was my cancer, it had spread, or "Metastasised", Just as I thought some weeks ago. He went on to say, the scans indicated that I have what looks like an extradural tumour in my spine, as the result of my cancer, spreading from the original site in my breast. It was his considered opinion, that it is inoperable, because it had fused itself to my spine, the best he could offer was a course of radiation therapy. He said it's just about as bad as it gets I'm afraid, and we must prepare for the worst. Remember when I told you, if a cancer spreads it's just about as bad as it gets, well, it was. When I asked him point blank what was the true, warts and all, chances of living more than a year or two, he tried to let me down gently he said with a course of radiotherapy, and a new drugs regime being the only course of treatment he could offer, being palliative, he went on to say I would feel o.k. for quite some time, but the latter stages would come on very quickly, and I would develop cachexia, rapid weight loss and muscle waisting. I said, just tell me please. He paused for a second, then said a year is about as much as I could hope for, and thought that eight or nine months was a more likely scenario.

Bloody hell! Just when I thought I had beat the cancer. He tried his best to instill a positive attitude in me and told me to go and enjoy life to the full. I think he meant doing all the things I wanted to do, but never got round to doing. Things like seeing Niagara Falls, or the Rockies, even, actually buying something from Harrods. Visiting the Rockies, or Niagara, just for the sake of it was a waste of money, And I probably wouldn't be able get insurance anyway, but I might just buy something from Harrods. My first thoughts, as with any mother, was for my family, not just my husband but my two sons, two of the best sons any mum could wish for, I was so proud and happy for them both, as I watched them mature into decent young men who will, I'm sure, make their mark in the world. Even though my future looks bleak indeed, I was just as worried about my husband's future, he had given up a good job some three years ago with the chances of getting promotion and a good retirement package. A similar job now is slim indeed and he was a long way from retiring. My "life" is, as an Member of Parliament, and I must try to and carry on, not only with my work in Parliament, even with the junior port folio on health, how ironic is that. I still had a lot to do, and a lust for life, and not forgetting my constituency work, but it was easier said than done.

A NICE LITTLE EARNER.

We began to think what was the best way forward. What would be the financial situation be if I resign, or should I resign, I feel o.k. for now. There were a lot of questions, non of which had answers at the moment. We thought it might be a good idea for my husband to spend some time in the house of commons library when it was quiet, to see what he could find in the small print of Parliamentary rules and regulations. When first becoming an M.P. we are given a publication called the "green book" this is a simple synopsis of the rules, but nobody really reads it more than they need, I'm sure I didn't. My husband was ferreting away in the library as often as he could without drawing to much attention to what he was researching, but he could easily say it was for me, well, it was! What he found was very interesting. There are five options for leaving Parliament. The first and most obvious is to simply step down when the next election is called. The second is to contest, but loose your seat at the next election. Third, commit a crime, serious enough to attract a harsh sentence, and hence a criminal record, or go bankrupt, in such circumstances you are barred from sitting as a Member of Parliament, and therefore your seat becomes vacant and a writ is moved for a by-election as you are effectively sacked. The fourth way is to resign your seat, but, a Member of Parliament is technically forbidden to resign, they must apply for the "Chiltern Hundreds" this is fictitious job for the crown, and as such, an M.P. cannot work for the crown, and continue an M.P. The fifth option, wait for it, is to die, and on the face of it the most lucrative, but the least attractive by a long way. Well, those are the options. Lets have a closer look. Option one, to step down, after all the hard work over the years, at school, at university, in the town hall, now in parliament, then to give it all up without a fight, not an option I like. Option two is to loose my seat at the next election, I feel pretty confident that I could win a second term, even with a small majority, so that's out. Option three, commit a crime, how much of a crime? or go bankrupt, I wouldn't have a clue where to start. Besides it would embarrass my husband and the boys no end, especially as they both graduated with flying colours, and are now at teacher training college. Option four, Resign, not so easy, asking what to do, to be Steward and Bailiff of the "Chiltern Hundreds", for the Crown. It all seems a bit messy, and besides we didn't want anyone to know my cancer had returned, just in case I change my mind. Option five. Die, Well, I'm odds on to die inside a year anyway. Truth be known, resigning, stepping down, committing a crime, which ever way you put it, non of them are a very lucrative option. The resettlement grant for the first three options, is about half of my basic salary. About thirty thousand pounds. Option four is rubbish, if a member wants to jack it in, tough, all I would get is out of pocket expenses to pay off staff, and other out of pocket exspences, and sod all for me. As we looked closer at all the possibilities, the best one is, believe it or not is to die as a sitting member of parliament. The stark reality is the fact that I am dying. So lets have a closer look at what can be a very nice little earner. When a member of parliament dies in office, the death in service gratuity, is a fantastic four times basic salary, at present rates it would amount to two hundred and thirty six thousand pounds, tax free. And as my husband is my employee he to would be eligible to some redundancy money as well, we didn't try to find out just how much, it might raise a few eyebrows, and set tongues wagging. Have you seen anyone raising their eyebrows and wagging their tongues, neither have I, but you know what I mean,to many questions could look a bit odd, people might think I was going to sack my husband and share the spoils, good idea, never thought of that one. All this may sound very clinical and measured, even callous to some people, l suppose it is really. As a biology teacher I know the odds of my living more than a year are not good. So lets think about it, I'm not long for this world, so if it's on a plate, grab a spoon, a big one. My soon to be widowed husband, and my two sons even though they're grown up now, left without a mother, well, wouldn't you try to make the best possible deal given the chance. We are now in late November and the nights were drawing in and things were a bit quieter on the political scene, so we had time to take stock of where we are, and plan what to do next. It is so frustrating to think that I had just got used to feeling well, and then to be hit by this bombshell, arrrrrrrgh! how cruel can life be. Is now the time to berate Him, or Her upstairs, as you do when things go wrong, but in my case disastrously wrong. This death in service gratuity, opens up all sorts of imponderables, when you're told that you have less than a year to live, and I have already used up quite a chunk of it, it really concentrates the mind. I did say earlier that I had got used to feeling well, and I had. But since the onset of my secondary tumour, I have to get used to the idea of, well, you know. This is where it may sound a bit cold and calculating, but it has to be done. It's late November, there's no need to call a General Election till spring 2006, but there is growing feeling it's just possibility that the prime minister could call a snap election as early as May next year, 2005. If I died between now and then, my husband would collect my death in service gratuity no problem! If the Government went the full five years, fair enough, I would be long gone, and my husband would collect big time, just the same. The one scenario that could scupper our plans is as follows. If the Prime minister calls an election for May next year, that's about seven months from now. I would most likely have to admit defeat, and step aside to let someone else campaign for my seat, and be satisfied with thirty thousand pounds, yes it's a lot of money but to loose out on two hundred thousand pounds on top of that, and tax free to boot. From now, to May, is only just over six months, and as someone famous once said, it could be a damn close run thing, if I was not well enough to be reselected to contest the seat, I would effectively give up a two hundred thousand pounds, not bloody likely. But this is a situation over which we had no control, well, not yet. We had to carry on as best we could, writing letters and e-mailing people, I get between fifty and a hundred letters a week, plus all the emails between colleagues depending if there is a particular issue to deal with. It's just as easy to deal with them from home, no one would know where I was, but we had to be careful not to draw attention to my state of health, the radiotherapy I'm having at the moment is a great help, but does nothing for my actual illness, it is just to help ease the pain, and the beginnings of weight loss and the sallow look that goes with it is something that no amount of pills and potions or radiotherapy cannot slow down, although I look o.k. at the moment, and don't feel too bad. I can remember the first time round, I felt ill from the chemotherapy but this time, I feel ill from the cancer, not a good sign. We felt we had to try to spend a little less time in the house of commons, but as the junior spokes person on health, I had to attend a number of briefings and meetings with all kinds of cancer related groups, a task I found more and more difficult to do. I did manage to persuade some of my colleagues that I was wanted elsewhere and became quite a passmaster at coming up with more and more ingenious excuses, especially when I had to go back home for my radiotherapy. The surprising thing was, very few people missed me, or if they did, they didn't say so. It's not unusual for me not to see my office colleague all week, and only a few M.P.s were actually in the chamber at any one time, unless there was an important debate, or a three line whip. When I mentioned this to my husband he said that he had come to that same conclusion recently and reminded me that there is no need to sign in or anything, there's no record of attendance at all, it's not like school he said laughingly, that was the first time I had heard him laugh for such a long time and it did my heart good to here him so. That night as we lay in bed, we began to hatch a plot from this embryo of an idea, with the lack of attendance scrutiny, no one bothering who's in and who's not, it's the only job I know of, where attendance is optional. When my husband had a closer look at this very strange anomaly of there being no requirement in law that says a Member must attend parliament at all, once you've taken the oath, and done the legal bits, no one bothers. So, as I had a session of radiotherapy coming up we thought that we would take a whole week off and see if anyone noticed weather or not we were in Parliament. We went home early on Thursday afternoon, arriving home about four o'clock. I couldn't help feeling like a school girl "wagging" of school, but that's being silly, we had some serious thinking to do. We decided to put the whole thing on the back burner till Monday and try to enjoy the weekend. The boys were away so we had a nice quiet and relaxing weekend. My radiotherapy session was set for Monday afternoon, and yes your right it was drizzling with rain, but it wasn't such a bad day. By Tuesday I felt a lot better. My poor long suffering husband did the cooking that night, it was the first time I ate a full meal since Saturday. During the rest of the week we answered what bit of mail there was and relaxed in the garden.

Monday morning we caught a slightly later train than usual. We didn't want to see anyone and get caught up in to much close scrutiny. We arrived at our little flat, down by the river just after noon. As my husband turned the key of our mail box, dozens of pieces of mail fell out all over the floor. Without thinking I bent down to pick them up and almost fell over, it was only the quick action of my husband that saved me from a nasty bump on the head. He gathered the mail from me and helped me upstairs to our flat and sat me on the bed. After a brew and a lie down I felt a lot better. A quick wash, and a bit of lippy, we went to the commons to see if we had been missed by anyone, But to our relief, no one, but no one had missed us at all, and no one said a word, it was incredible. We just carried on as if nothing had happened. However there was a feeling that if we were not missed, we were not valued, I couldn't quite get my head round that one, and I couldn't ask any of my colleagues had they missed me, it would seem very odd. We tried it again a few weeks later and guess what, we were not missed again except by the chief whip, he just said he wanted to speak to me the other day, and I should do my best to let him know when I was going to be absent. One or two of the library staff said they hadn't see my husband for a day of two because he had been spending a lot of time in there recently, but they were not a major concern for us. Everything seemed to be falling into place. If I was becoming to ill to attend parliament we could probably get away with it, certainly for a few weeks at least, that would bring us well into the new year.

The most awkward time I had was when I was asked to attend the "new intake" dinner in mid December, this would be my fourth. It's given by the party leader each year for the new members, these are members that entered Parliament at the last general election. We have, in the past gone out to a posh restaurant, but I think funds are a bit tight at the moment. So we dined in house as it were, in one of the many committee rooms near our leader's office. There were twelve of us, plus the leader, and a couple of secretaries at the dinner. Partners were not invited because not all the members had partners here in London so it would be unfair on them. Any new members at the next election, will, in turn, become the new intake, and we move up the pecking order and so on. The radiotherapy was a great help, but I was still a bit apprehensive about going to the dinner even though I didn't feel too bad in myself, and managed to put on a brave face. I've been wearing padded bra's as you know since my mastectomy, so at least I still looked quite "busty" if you know what I mean, so with a wool frock and a silk scarf and a bit more makeup than usual I looked, and felt better than I have for a while. I was somewhat relieved when I was seated at one end of the long narrow table where the light was a little subdued, and the members sat near me were of their own clique, and not in day to day contact with me. My office companion sat opposite me, but we hardly spoke as she was too busy cosying up to the leader all evening. I managed the cream of celery soup, but didn't eat the roll, as I wanted to be able to eat the main course. I was hoping it was something I liked, and it was, but there were just two components missing, yes that's' right, curry sauce and a bowl of fried rice. The chicken was very nice, with a rich creamy sauce. I'm glad I managed to eat it all, people do notice if you leave half of your dinner. As the evening came to a close I made my excuses and left as reasonably early as possible. As I sat in the taxi on my way back to the flat, I was relieved that no one said anything about how I looked, even if they thought I didn't look to bright, even my friend with whom I had shared an office for the best part of four years didn't say anything to me, only the usual shop talk. But more importantly, I was trying to listen in as much as I could, to try to gauge what the thinking was about the timing of the next election, but no one seemed to know very much. I'm glad I went to that "intake dinner", because if nothing else it proved a point that no one said anything about my absence. It really helped me to make my mind up. We were going to go for it! I couldn't wait to tell my husband. I arrived back at the flat just before ten o'clock. I paid the taxi driver and as I turned round, I could see my husband was waiting for me by the front door, he couldn't wait to here how I got on, and most importantly did anyone say anything about my appearance. Even though I was looking a bit pale and my hair was loosing it's body, no one said a word, it was great! We both felt that a plan was coming together. We watched a bit the tele' for a while, and went to bed feeling a lot more hopeful but at the same time somewhat circumspect. It was nearly midnight when we tured in and slept as sound as could be.

The next day, Wednesday was quite a busy day for which I was grateful as the time just flew by and I forgot my troubles for a while, we even went in the member's bar for a drink before we went home to our flat, we both felt quite good with the world. Christmas wasn't very far off and we were both looking forward to the break, I still had some letters to tidy up, and clear my desk, but as I was about to leave the office, I had a really awful dizzy feeling and nearly fell to the floor, I managed to get to my desk and compose myself just in case anyone came in. After a few minutes I phoned my husband to come up to the office right away as I didn't feel very well. I told him not run as he might draw attention to himself. As he rushed through the door, the look at me and sat down breathless, after me telling him not to run, he said he couldn't help himself. He said I looked awful, with an almost grey pallor, not quite what a girl wants to hear, but I did feel awful, and to afraid to look in my mirror. After about an hour and a couple of cups of tea, and one of my morphine tablets, my husband said I looked a bit better, by now it was getting on for five o'clock, so we went back home to our flat, well it's home whilst we're in London. As he helped me up the stairs I had the sudden realisation of just how weak my legs were. That dizzy spell really knocked me for six. After a sit down, he made a bite to eat and brewed up, we just sat for while, thinking we'd just had a lucky escape. Suppose someone had come into the office, it could have been very awkward for us. About seven o'clock he helped me to get in bed with a comforting hot water bottle and a goodnight kiss. The next day I woke up to broad daylight, as I gathered myself together, there he was, fast asleep on the settee. As I stirred to get up, he heard me and asked how I was feeling, ok I said, as we both looked at the clock at the same time, and laughed, it was quarter to ten, but it didn't matter, no one's bothered or even cares what time we get to the commons, unless we were to attend a meeting, or giving a presentation. I couldn't eat anything, but my husband had a hearty breakfast of two eggs, the remainder of the bacon and four slices of bread and butter. The rest of the bread he threw out for the birds. He wasn't being greedy as such, he said he wanted to finish off any food that might go off. Chance would be a fine thing I thought as he wiped round the fridge with some anti-bacterial cleaner to stop any mould growing. What he had in mind, and to my delight, he said we're going home this afternoon, you need a proper rest. After a quick shower, with my husband on standby near the door in case I felt dizzy again, or needed any help. Once

I felt I was o.k. and was able to towel myself dry, my husband felt confident enough to leave me for a while, he dashed off to the commons. Half an hour later I was refreshed, and fully recovered from what felt like a very bad dream. Feeling, and looking better, I'm sure would please my husband, and make him feel better seeing me more like my old self. Whilst he was away at the Commons, I packed a few clothes emptied the laundry basket as I usually do when we go home, and have a tidy round. I would hate to come back to an untidy flat.

After little more than an hour and a half I heard my husband bouncing up the stairs like a ten year old. He said all went well, just a couple of nods here and there and keep moving. I must say he's getting good at this. We sat down for a while, gathered our thoughts over a cup of tea, then I rang for a taxi to take us to the station for the one thirty train, and it's home James and don't spare the horses. The journey home was not the same. As I sat watching the world whizz by, I usually read up on constituents mail, or write a few notes. But on this occasion, I felt as if I was running away. After all the work, all the foot slogging, the cancer, the despair as we sat in that little green waiting room, the joy when I was given the all clear, and that little green room lit up as if by the heavens. The euphoria the first time I won my council seat, then the double euphoria when I took my seat in Parliament. I never thought in my wildest dreams, I could ever feel so down hearted as I am at this moment. Time for a sticky bun me thinks, I was feeling a little hungry not having had any breakfast. As we watched the world pass by on this dull day in December, I wasn't looking forward to Christmas one little bit. We used to love dressing the tree, and watching the boys in the school nativity play. They each played one of the three wise men. Then there was wrapping the kids presents, two of everything, even though they're not identical twins they have similar tastes as some of you with twins will know. It was dark when we arrived home just after four o'clock that Thursday evening. Thank heavens for central heating and timer switches. As we pushed the front door open we could feel the heat on our faces. The whole house was lovely and warm. You know by now what the first job is don't you. After which we unpacked, I shoved the laundry in the washer, fifty degree wash and bobs your uncle. As we sat by the pseudo log gas fire thingy watching the same repetition of the gas flames. We both dosed off for while, even thou' the washer was sloshing and pumping in and out. It must have been the central heating. After tea, my husband put the washing in the dryer in the garage then made a brew and sat down beside me. I had to remind him about this unique position we were in, cold and calculating as it may look. My poor husband was reluctant to discus the financial side of things, as it all seamed so mercenary, but I told him in no uncertain terms that if I was going to die sooner rather than later, we might as well get the best deal out of the situation as we can, a deal that could net us the best part of a quarter of a million pounds. People say I must be positive, well what can be more positive than securing a shed load of money for my family. This was a positive I could deal with, simply because I had no choice. The problem was, keeping my husband focused on the big picture, and the date of the next general election. I felt able to carry on for a good few months yet, but the timing was crucial. It was just possible that the election would be called in the spring of 2005, probably in May. The reason for my thinking on this, was the fact that the war in Iraq was looking more and more difficult to justify, and the death toll of British service personnel was fast approaching one hundred, a backlash was always possible as a protest against the war and the government. If the prime minister delayed the expected election for another year it could have disastrous consequences for the Governments re-election chances. You don't need me to tell you that from now till May is just under six months, my status now, is as an "active" member of Parliament so if I die now I can still collect. But my status would change to that of a "degraded" member during the election campaign, but even then I would still collect. The problem is, if I win the seat, I become a re-elected member. I will have to go down to Westminster to take the oath and sign the test roll, in order to become an "active" member once again. It's simple, if I can't take the oath, for my second term, my husband gets nothing, end of ! Therefore, in order to be eligible for death in service gratuity I need to become an "Active" Member as soon as possible after winning seat. I couldn't imagine what a bitter blow it would be, after all the work we've done. We just have to maintain my status as an "active Member at all costs to maximise our chances of collecting the grand prize. All this may sound very unseemly to you, but it's quiet simple. The upshot is, if the Prime minister doesn't call an election, and I die as a sitting member, then my husband and two sons will benefit enormously, but if, as I dread, he calls an election for May next year, and I'm not quite strong enough to contest the seat, I would suffer a double whammy, being forced to admit being defeated by this cancer that I have been fighting for the past nine years, and step aside from the contest for some upstart to benefit from all my hard work. And for me to end up with a much reduced resettlement grant when we could pocket an extra two hundred thousand smackers. What a dilemma. But I see a glimmer of hope, if I could convince people that I was up to the job without them wondering why I would want to say such a thing in the first place, we might get away with campaigning for the seat again, but we would have to box clever. It will need a lot of careful thought and a lot more luck. My husband had made his mind up to attend my constituency surgery on the Saturday morning, and do the shopping afterwards. All I had to do was make a list for him of likely problems at my, sorry his, surgery, and write a shopping list for him. When he came home about one o'clock, I had some dinner on the go, well on the go, is pushing it. Cheese on toast was the best I could manage, but he said it was scrumptious and particularly liked the t.l.c. As he ate his toasted cheese he said only one elderly man came to see you about his council tax. I listened to him with a certain sympathy and gave him the name and number of his local councillor and said he was better placed to deal with such an issue. Afterwards he just had to try on his wellies again to sweep up more leaves. I said to him, if he wanted, he could take his wellies to London next week. He couldn't help but throw some leaves over me, it's a good job they were dry. As we paused for a moment, my mind went back to the time when I finished my first Marathon, when I threw water all over him. It seams like a hundred years ago. With that, I went indoors and had a little weep to myself. My husband, sensitive soul that he is, left me alone and carried on with his very own sweep-a-thon. We decided to go down to Westminster on Thursday rather that Wednesday even though we would be on our way back home before we knew it. The reason for not going on Wednesday is because of Prime Ministers questions when the house is usually quiet busy. So we wrapped our selves up in our own little world for three lovely days.

The journey down on Thursday was long and slow, with the clickerty click, and no sticky buns, that made me feel even more grumpy. On our way to our flat I cajoled my poor husband into asking the taxi driver to find a sticky bun shop for me. After some chunnering he did eventually pull up outside an old fashioned bakers, he said the missus usually shops there. To my annoyance, the shop was just around the corner from our flat, and we didn't know it. With the usual cup of tea, those sticky buns were scrumptious. My husband said they must have cost about five pounds each, with the amount the taxi driver charged going round the one way system, and worth every penny I said. As we sat in our little bed sit', we ate about ten pounds worth of buns, according to his lordship. The Christmas recess was almost upon us, and it couldn't come soon enough. It's a good job we went back when we did. On Fryday afternoon we were expected to attend, along with the rest of my colleagues, a bit of a do in the leader's nice plush office for a Christmas drink, and listen to him telling us what we already knew, about how we can do better, how we must do more to hold the Government to account, then went on to wish everyone a happy Christmas and a prosperous new year. We didn't want to be seen going off to early, as we might be missed, but as soon as one or two started to make their way to the door, we managed to sneak off a bit sharpish and hope no one noticed. We made a flying visit to our flat as we do before we go home. We had only spent one night there, so it was grab the washing and a few bits and pieces and off to the station. The journey home is beginning to get very tedious now, and I really am looking forward to a nice break. When we arrived home that Friday evening I was beginning to feel sick again, I called my husband to come down as he was upstairs unpacking, I told him how I was feeling. Have a nice hot soak in the bath, then you can have a lie down for a while before tea, shout if you need any help. My husband was the most understanding man I have ever known, that in itself was a great comfort to me as I can feel the cancer beginning to take a hold. We didn't do any Christmas shopping as such, we just spent the week doing nothing much, just looking forward to the boys coming home next week to spend Christmas and new year with us. I had just finished getting dressed and brushed my hair, whats left it, when I could hear the commotion down stairs that heralded the arrival of my boys, I went down to see them, how good it is to see them looking so fit and healthy. It was difficult to keep them from giving their mum the usual big hug, but there was nothing I could do about it. I got the sense that they were somewhat taken aback when they realised how much weight I'd lost. But as ever, there dad managed to steer them away from my problem by asking them as many questions as he could think of, in particular about any girls friends, they both laughed and said they had been far to busy for girl friends, I thought, I hope they like girls. After a catch up, and something to eat, I felt a lot better. I told the boys I had 'phoned their gran' the other day, to ask her to come and stay with us for Christmas, but she said she liked her own bed to much, and didn't want to be a nuisance. We all agreed that I should ring her back that very evening, and make arrangement to go up to Kendal in the morning, weather permitting of course, it can be a bit wild up there in December. About eight o'clock they went for a drink and a game of pool with their dad. The pub was only a few minutes walk away so they could have a couple of pints. They asked me to go with them, but I was quite happy to watch a bit of tele' and have a dose on the settee. About ten o'clock, I was disturbed by the sound of the men coming up the gravelly drive. The kettle was on as they opened the front door still arguing about who won that last game of pool, it was so good to hear them enjoying themselves so much. I told the boys what their gran said when I rang her, she would be delighted to see us, but more importantly her two little boys, all six foot of them. After a catch up of the news on the tele', it was soon time for bed as we wanted to get an early start.

IT NEVER RAINS

What a lovely sunny Thursday morning for a ride up to the lake district. We were well on our way up the M.6, looking at the countryside and feeling quite at one with the world. Until my mobile phone rang, who can it be at this time of day, I nearly didn't answer. It's a good job I did, it was my mother's neighbour, she rang me at home, and was redirected to my mobile 'phone. She said she called at my mother's house this morning as usual about nine thirty to see if she wanted anything from the shops, but couldn't get an answer, then she tried her spare key in the lock, but it was bolted on the inside. As she looked through the window, she could see my mother lying there on the floor. Even though the neighbour was obviously upset, she did manage to explain what she thought had happened. All she could do was call the police and ambulance service, the police forced the door open, but it was to late to save her. The paramedics said it looked as if she had had a heart attack during the night, and died where she lay. As you can imagine I was very quiet as I listened to the message from the caller, my husband said whats up, somebody died. When I said it was my mother, he was really upset for saying what he said, but I forgave him. How could he possibly know who was on the other end of the 'phone. I told the neighbour we were already on our way to see mum, and would be there in about an hour. My mother dying so sudden was the last thing anyone expected, she seemed o.k. the last time we saw her about two months ago, and sounded o.k. on the 'phone last night. Upset as I was, we had to carry on, and wondered what to expect, we had to find out what we should do. We were met by the neighbour who, unknowingly to us, had cared for my mother more than we realised, for which we were very grateful, and told her so. She then went on to say that my mother had not been too well for these past few weeks, and had had the doctor call on her more or less every day, but wouldn't allow him to call us. Then we spoke to the waiting policeman. He told me my mother had been taken to the mortuary at the local Hospital. Then he explained to us that a C.I.D. officer had had a look round, just make sure there was no signs of foul play. They looked in the cupboards and bathroom cabinets to see if she had any medicines that could have contributed to her demise. He quickly assured us that after a word with the paramedics, they were satisfied that she died of natural causes. He was one of the old school coppers and knew how to deal with situations such as this, and gave us time to look round the house. Whilst my husband was upstairs, I did a very stupid thing, I asked the policeman where my mothers body was found, as he pointed to the floor just in front of the fireplace, I was sorry I asked, I looked down at the floor and could see a damp patch where she had wet herself as her bladder had drained after she died. I thought to myself, how undignified it was to be found dead with wet knickers.

The policeman said she had a duster in her hand,

it seems as though she was trying to dust round before your visit. I told my husband what had happened, and needed to go outside for a minute. That's ok, go and sit in the car with the boys for a while, like I keep telling you, a real rock. After twenty minutes or so, I went back inside. He said he had given all the frozen food from the freezer, and the bits from the fridge to the neighbours, then he sprinkled copious amounts of bleach inside the fridge and freezer, tilted them back to stop the water dripping out. He read the meters as he switched off the electricity, gas, and water, and left the water taps open. The last thing we needed was a flooded house to clear out. The neighbour said she would keep an eye on things, and let the land lord know what's happened, and tell him about the door being forced open. It was at this point, I think I did something that was sort of snobbish, a sort of euphemistic pat on the head, I gave the neighbour fifty pounds and told her to buy something nice for herself. I didn't tell her it was the fifty pounds I was going to give to my mother anyway. The policeman said we would need to identify the body, and suggested that the sooner we did, the easier it would be. He said he would need a statement from us to keep things "tidy" before we went back home, he would get a statement from the neighbour tomorrow, as she's to upset at the moment. The policeman said it would be easier if we were to follow him to the hospital so we wouldn't get lost. We were there in five minutes. My husband said you and the boys can stay in the car. I'll go and identify her, and sign any papers he said, but I just couldn't walk away and not see my mum one last time. The boys felt the same way too, so we all went to say our goodbyes. Whilst we were in the chapel of rest my husband went to collect the medical certificate from the doctor to give to the registrar. A man of the cloth with another couple came in and offered his condolences and said a little prayer for the dearly departed. He introduced himself as the Hospital Chaplain, and said he had been told about our situation. I asked him which church he represented, he said he was one of many members of the clergy from Kendal Parish church. I found out later that it is one of the biggest parish churches in the country, and that the cemetery was closed because constant flooding. I looked up to the heavens for a minute, I thought we would have needed to contact someone about my mum's funeral soon. It makes you wonder. After a few minutes chat he said he could act as a go-between and help with my mum's funeral arrangements. He gave me his card, and believe it or not, it had the churches web site and his email, address, he said he could put us in touch with a Funeral Director to see to things if we wish as we lived so far away. We shook hands thanked him for his kindness, and said we would be in touch with him very soon and would most likely go along with his suggestions. And that was that. I'm glad it was all so quick, I was so close to tears and held onto my husbands arm as we left the chapel of rest. We were then ushered to the relatives suite to give a statement to the policeman. He said he wanted me to fill in a few blanks for his report, things like how her general health has been, and what medication was she taking. I'm ashamed to say, but I couldn't hold myself together any longer, I just had to get out of there. That was the cue my boys had been waiting for, as they said we'll come with you mum. Once again my poor husband had to stay and pick up the pieces. When he came out to the car, he said because he didn't have all the answers, the best he could suggest was that they speak to her G.P. The policeman gave him his card and asked us to phone him if we thought of anything he should know of, or if we had any questions. He would get her doctors name from the neighbour tomorrow morning and any details to help complete the formalities. He went on to say that as tomorrow is Christmas eve, we would not be expected to register the death till after Christmas. He said he would inform the registrar of my situation, as it may be difficult to get back to Kendal within the five days as required. There would be no need for a Post Mortem, as my mother had been under the doctor's care for quite some time. I was glad, as that was one less trauma to deal with. I closed my eyes and wished we had a supersonic car to get us home. We arrived home about nine o'clock, starving hungry, dying for the toilet, and knackered. It's not yet sunk in just what a bloody awful day it's been. My second move was to make some supper, you know the first move by now, the boys and my husband were starving . Me! I was off to bed with some cocoa for a good nights sleep.

Christmas eve, we used to be running around doing last minute things like you do, but not after yesterday. We had a very non day really, all of us. Christmas morning I felt wretched. But as the local Member of Parliament, I was expected to go to the Christmas morning service as I usually did, I suppose it serves my right for being such a hypocrite going to church in the past, just to keep well in with the voters. Although I was not feeling at all well, and far from in the mood, I felt I had to make one last effort. My Oncologist gave me a sightly stronger painkillers to be taken when the pain was most severe, well, I did feel bloody awful. With my woolly frock, and winter coat, we made it to the service, with my husband, and our two sons. I really shouldn't say this, but the service was, well, meaningless to me, why would a loving God to whom we have just been singing our praises, put me, and my family through this living hell, I mean my cancer that started some nine years ago, and put me through a time when I thought I was going to die, a time when I thought I was going to live, a time when I didn't know what to believe, a time when I was given hope, a time when hope was lost, not only for me, but my husband, and in some measure, our sons too. Was it really a man or a woman up there, or some other Deity, or even a monster ? As my husband drove us home I thought back to the times we would call in our local hostelry to wish everyone a Happy Christmas and a prosperous new year, then dash home for Christmas dinner, afterwards we would all sit round the television to watch the queen's speech as we opened our presents to each other. But not this year, can you imagine just how gut wrenching it must be for my husband, and the boys for that matter, to try to find a present for someone who is dying, specially if that someone is your wife, or your mother. A pair of slippers, a cardigan, a very thin book. The boys said they had come up with an idea, and put it to their dad when they were out together the other night, but I would have to wait 'till the new year. I hope it's not too long to wait I thought to myself. Dad and I bought them a beautiful Rolex watch each, inscribed "Nil desperandum Christmas 2004"

When we arrived home I went straight to bed for lie down, I couldn't be bothered to watch the Queen's speech, and in my heart of hearts I found it difficult not to think of the future, and how next Christmas might be, the potency of the painkillers is amazing, after taking just one, I fell into a dreamy asleep. I dreamt I was in the Speaker's chair, wearing his wig and gown, like they used to do some years ago, and bossing everyone about, but the only words I can recall saying were "order,"order"," The next thing I knew, was my husband shaking my shoulder, as he sat on the edge of the bed and gave me a cup of tea. Then, curiously, he said he thought I called him "the honourable gentleman", how silly is that, he's my "honourable husband". By now it was nearly seven o'clock. I couldn't be bothered to get dressed so I put on my husband's big blue dressing gown, it was much cosier that my silly fashionable pink fluffy thing I used to wear. I went down to join him and the boys, but I didn't feel like any Christmas dinner as such, but I was feeling a little peckish, so one of the boys made me the best turkey sandwich ever, and of course another cup of tea. Christmas evening on the tele' was the usual mix of Christmas with this celeb' or that celeb' we did manage to find a film, some crime drama, but after about twenty minutes we all agreed we were not in the mood, so we had a cup of tea and got out the good old standby, scrabble. It was the first, and, I thought to myself, probably the last time we would all play together as a family. After the death of my mother we were all a bit somber and down in the dumps. The following day, boxing day was Sunday, and as it was the first time that Christmas day was on Saturday and boxing day on Sunday since I became an M.P. I wasn't sure if I should attend church for a second time, as it was Christmas Sunday, but my husband thought I had done my duty for now, and I was glad. I felt we must go down to Westminster after the Christmas recess, if my legs would carry me. My legs were beginning to feel a lot less robust than they were only a few months ago. Legs that turned my husband's head all those years ago when I used to play netball, legs that carried my babies for nine months, legs that pushed the pram with my most treasured possessions in it, legs that carried me through two marathons, legs that trudged the streets round my constituency, legs that, well, I must stop now or I'll start to cry. But alas, as I run my hands up and down my thighs, I can feel that feeling of wasting away that one gets with the onset of the final stages of cancer. I didn't say anything to my husband, but sometimes, I wish it was all over. The new year was upon us all too soon. My mother's death two days before Christmas was the most awkward time for us. Yes I know it sounds a bit callous, but we needed to sort out most of the formalities within five days. My husband said we would be allowed a bit of leeway as it so near to the Christmas period, and the distances involved. He went up to Kendal on the first Wednesday after Christmas to sort everything out, and I mean, everything. He went to register the death, and get some copies for the various formalities such as the department of work pensions, the post office, insurance, and a couple of spare. Then he went to the benefits office, the post office to redirect her mail, contact the gas, electric, and council tax office, and God knows who else. But when he eventually got home about six o'clock, he said thank goodness that's all done. Well not quite , we had the funeral to attend. The chaplain we met at the Hospital was very good, he acted as go-between us and the funeral director. We couldn't have asked for anyone better. Because of the delay, the funeral was to take place in the Chapel at the local cemetery on Tuesday morning at eleven o'clock on the fourth of January. What a way to start the new year. We went to collect the neighbour who had been so good to my mother, then we followed the hearse the mile and a half to the cemetery. The boys asked if they could be pallbearers for their dear gran with the two men from the funeral directors, a gesture that pleased me more than I can say. With a few neighbours, we filed in behind the coffin and the Vicar into that little Chapel on a cold and damp day in January. I don't recall what hymns we sang or quite what was said, it was all to difficult for me. But the one thing I do remember, strange as it may seem. The plot of ground where my father, and now my mother are to lie together for all eternity was, even on this dull January day, bathed in sunlight as if my farther was opening the pearly gates for his long lost love. A strange shiver came over me, then, as the coffin was lowered into the ground I was overcome by a feeling of piece I never thought possible. I stood there for few minutes and said a little prayer. We thanked the Vicar and everyone for there kindness at this difficult time. My husband had a word with the undertakers, and no doubt thanked them for a job well done. The journey home was very quiet. You can only say, she had a good innings, so many times. I suppose seventy nine isn't so bad. It's a damn sight more than I can expect. The boys didn't have time to come home to have anything to eat, they had a train to catch and couldn't miss it as they should have gone back on Sunday. They had to get back to their respective grind stones. We waved goodbye as the train pulled out of the station, and with a tear in my eye, I had the distinct feeling I wouldn't see much more of them, maybe Easter, if I'm still hear. When we got back home at about five o'clock, there was a real chill in the air. After our usual cuppa' we talked about my mother's estate, such as it is. We went up to my mother's on Sunday after the funeral, I didn't want to go, but my husband said I needed to get out of the house for a few hours, and thought the roads would be quieter, and they were. We went to clear out her personal things like her laundry basket and other clothes that were of no use to anyone. So, apart from a couple of boxes full of papers, there was little left to clear, just some old furniture, some even had the post war utility marks stencilled on the back, but nothing for the antiques road show, or even cash in the attic, more like scrap heap challenge, and the carpets had seen better days. Sorry mum. We got the name of a house clearance firm out of the local yellow pages, courtesy of the neighbour. She offered to let them in and see to things for us. As we sat in her little cosy front parlour with a very welcome cup of tea, we gave her all the keys we could find so she could give them to the land lord after the house was cleared. We told her she could have anything she wanted before the place was cleared out, they were to shift everything that was left. £50 was all we got from the house clearance people for a life time of living. I told them to give it to the neighbour and thank her for her trouble. They take pot luck if any of it can make a few bob, makes you think, but it's less painful to let someone else do it. We came away with a number of old albums, a battered Oxo tin of old of photographs, boxes of papers, and yet more photographs in shoe boxes, a few bits of cheap jewellery, dads gold fob watch, and some odd cuff-links and even a pair of collar studs and some loose collars, I'm sure they will amuse the boys no end. We didn't tell them we had the boxes to go through, they had had enough to put up with over Christmas and new year. We had a 'phone call from my mother's neighbour one afternoon a few days after we left. When I heard her voice, it all came back to me, just as I was beginning to get back to some semblance of normality. She 'phoned me to say everything was cleared out, and thanked me for the £50, she said she would give it to cancer research, how's that for irony. She went on to say she would miss my mother very much, as they had been good friends for many a year. I thanked her once more for her kindness as she said goodbye.

Going through my mother's things was a task I didn't relish one bit, but the sooner we get on with it the better. As we sat by the fire on a dull rainy January afternoon we prepared ourselves for a journey down memory lane as we uncovered a collection of photo's from the war, their wedding, me in my cot, our wedding, the twins in short pants on holiday at Grange over sands. Aunts and uncles long forgotten, insurance policies, newspaper cutting of me finishing both marathons, becoming an M.P. It was all beginning to get to me, then my husband came up with a good idea, why don't we take the insurance policies out, and put the boxes to one side for now, and take the lot to our flat in London, it's a completely different environment and it could be less emotional for you. Good idea I said, do you think it would be wrong to celebrate with a chicken curry, not at all he said, what time would you like it served madam, about eight o'clock please sir, very good. We both laughed out loud, and hugged each other. With the evening creeping up on us, we just sat and stared at the gas fire's flames blobbing up and down, on my side, three short blobs and a pause, on his side, three long blobs, and a pause, and mused over what a bloody awful Christmas we've just had. With my mothers unexpected death, the funeral, clearing out mother's bits and pieces and all that it entails. Like I said, a bloody awful Christmas. Within a three week period, my mother had died, Christmas had come and gone, my emotions had been up and down like a yo yo, my mother was laid to rest with my dad in Kendal cemetery, her house was cleared out, redecorated, and new tenants about to move in. It was as if my mother never existed. It's that thought that makes me feel so sad. The only little bit of comfort I could glean from this awful situation, was the fact that she wouldn't be around to witness my demise. Well, that's enough about my mother. My worst thoughts looking back over this Christmas, bad as it was, is the sad fact that it is almost certain to be my last. At six o'clock we watched the news, just make us even more depressed than we were already. Chicken curry with fried rice, scrumptious, need I say more. The rest of the week was cold and snowy, I did a bit of house work, looked at a few papers from my health portfolio. My husband made several attempts to vacuum his car, but it was to wet and cold. It's funny how a man can have a face like smacked bottom just because they cannot get their own way. Surely no one expected me to hold a surgery so soon in the new year. If they did, they would be disappointed. Silly me I forgot that old fashioned thing called a telephone. The first Saturday I have had off in ages, but as people realised I wasn't there, the 'phone started to ring, then ring again, no sooner had I put it down, it rang again. Not one of them was important, yes I know I'm a member or Parliament, but give me a brake. Most Saturdays only two or three people come to my surgery. As I watched the first P.M.Q's of the new session on television, I felt so frustrated, and said to my husband that we really must make an effort to go down to Westminster just to show our faces as it were. He couldn't help but agree with me, I said the best time to go would be a day or two after my radiotherapy next Friday. So Monday looks good to me if I feel up to it. Do you remember when we went to the garden centre and he bought his wellies, well, we went back on Friday, this, to most people might seam a bit morbid, but when you think about it, lots of famous people do it, yes I know I'm not famous. When he and the boys were having a drink and a game of pool before Christmas, they came up with the idea of a tree. My husband wanted to plant a tree as a sort of Memorial, before I go, as it were. Is that a bit too morbid.

MALUS DOMESTICA

We had a look round the garden centre and found a lovely apple tree, well not the tree itself but a very nice picture of one we liked, it was Malus Domestica, the man in the centre told us all about it, how nice the white, pink edged blossom would be, and the apples are delicious in the late summer. My husband could see I was beginning to fill up and told the garden chappy we were running late and can he order one now and pay for it. Of course you can sir, It'll take four to six weeks, for it to come through, if that's ok. I'll just take your order and you can get off. After a few minutes, we were on our way home. Whilst it was a very nice thing for them to do, maybe it is a bit morbid after all. But like I said what else could they get me. Remember, we still have a vested interest in the timing of the next general election, the later the better for us, ching ching, we could do with another outbreak of foot and mouth, for purely selfish reasons you understand. If the election was called at just the wrong time it might be awkward at constituency level. But thought it prudent to say nothing unless we absolutely had to. It was Monday morning, the end of January. I felt fine after my session of radiotherapy on Friday with two days to get over it. We decided to catch a later train to London, about ten twenty something, we wanted to miss not only the crowds, but we didn't want to bump into anyone we knew, and be sat near them for to long trying to look my usual self, in good health and good spirits. But as it happened it was an uneventfully journey down. We took my mother's papers with us in a small brown case, but will we ever get round to looking at them, I don't know. The insurance by the way is only worth about £450. When we get it. I suppose it will help towards the funeral costs. We got off the train pretty damn quick, and made a bee line for a taxi, we keep clear of the underground, I don't want an elbow in the ribs. We went straight to our flat to gather our thoughts for a hour or so. After a brew, and a look at the one o'clock news we took a taxi to the Commons. We used to walk most morning come rain or shine, but once again this was one thing I couldn't do any more. It's funny how such little things rise up as being more significant or important than you could ever imagine, not just the walk to Westminster but the smells and the noise of a busy city with the couriers on their push bikes, and often are more of a hazard than the usual traffic. The tourists are a feast for the eyes, in a myriad of colours, shapes, sizes, and yes, even smells, with the sound of so many different languages echoing from around the world, full of vigour and vitality. As we sat in the taxi waiting for the lights to change, just for a few moments I forgot my cancer and just lived a little more. As we approached security at the entrances to Westminster I was getting a bit apprehensive because quite a queue was forming as all the members passes were scrutinised, this is because the personnel had been changed by the security people as a precaution, it seems that familiarity breads contempt. And Eureka! there in lies the key, as two would become one familiar face, it figures that if my husband was seen at Westminster, people would naturally assume I was there too. This was just the break we were looking for. A cleaver trick, a double act, and one that would prove very useful in a few months time. I went up to my office and was surprised to see my friend there I don't know why I was surprised. She was busying away at her computer sending emails and such. Hello, she said, had a nice Christmas? Quiet, with the family you know, I didn't mention that my mother had died, I didn't want her feeling sorry and hugging me. Then she jumped up from her desk and said, must dash, lots to do and all that, and off she scurried down the corridor. I got the distinct feeling that she wasn't exactly avoiding me, but didn't want to make eye contact with me, and not a word about my absence these last couple of weeks. As my husband went through the mail, I read the "whip" that's a sort of letter or summery of the weeks events, from the chief whips office, and a note of any three line whips, that's a very important vote that we must attend, and vote with the party. A requirement that is literally under lined three times. Whilst my husband was going through the mail, there comes another eureka moment, I thought, would anyone dare to ask a nun if she was a virgin? or how well, or how long have you got? to someone from suffering terminal illness? you wouldn't dream of it! It didn't take long to tidy my desk, as my husband finished sorting through the mail then made us a cup of tea. It was great being back in the old routine, I felt so good, I did wonder if the doctors had got it wrong. Once we were finished, I was itching to go in the chamber, but my husband wasn't so sure, he said, just stand at the entrance to the chamber and see how you feel, as we approached, I had a tremendous adrenalin rush and felt that maybe I could sit in for a while, till my biology teacher knowledge kicked in and I realised that this adrenalin rush would soon die down, and could leave me weaker than I am right now. I thought to myself, who are you kidding, it would be folly of the worst kind. I couldn't help but feel sorry for myself and so frustrated as we stood there wondering what our next move should be. As it was nearing four o'clock, we decided to go back to the flat and rethink our strategy over hot curry, I love hot curries, even though my doctor says I shouldn't, but they're one of the few meals I can really taste now. The last thing expected to loose, was my appetite, but these things happen. The next morning, Tuesday, I'm glad my husband had a good nights sleep, because I certainly didn't. He went to the commons nice and early to collect my mail, and pick up any messages. When he got back to the flat, he took one look at me, and said I didn't look well, even though I managed to shower and dress myself, and put on a bit of makeup. We decided not to go back to the Commons and risk anyone seeing me like this or saying anything that might put us in an awkward position. My husband was seen by a number of people, including the Party's chief whip, he asked how I was getting on, and said he had not seen me for some time , and would like to have a word about my future roll as junior health spokesperson. Shit! I thought, what do I tell him, he'll assume I'm here, if you are. Then my husband smiled, I told him you had gone to be fitted for a some new padded bras', that embarrassed him, so he shut up and scooted off down the corridor. Very cleaver, I said, we could use that one again. As we still had an eye on the main chance, we elected to go home and keep our powder dry as it were, and caught the noon train. There were few people travelling north at that time of day and to our relief, the journey was uneventful. We arrived home at about three thirty. What a relief to sit in a nice comfy chair and put my feet up, and yes, a nice cup of tea. I found it increasingly difficult to stay in London. We came up with the idea that if my husband went down to London on a regular basis, say, two, sometimes three times a week, on a slightly later train so as to avoid bumping into anyone he knew, might be the best way forward, because as I said earlier, see one of us, you subconsciously see both of us. He could go to the commons to check my usually full post box, collect the Whip, and then go straight to our little flat to make sure we hadn't been burgled or the like, though there wasn't much worth taking. Some of my letters he could answer himself, in fact he was perfectly capable of answering nearly all of them, I used to be a teacher you know! But we had to have some semblance of normality, and self worth for me, so he brought most of it home for me to deal with, including the "whip" you know the letter of weekly events. I'd flick through it, then put it to one side to read later, but I never did. The Monday following my husband's chat with the chief whip, there was a personal letter from the man himself, delivered by Royal mail to our house, rather than to my office in the commons, that in itself was of concern to me, and I didn't like it. He asked how I was getting on as he hadn't seen me for a while, then to my relief, he made a very awkward situation easy for me. After saying he was aware that I was quite ill, he went on to say he presumed I was finding it difficult to fulfil my duties in the house, with particular reference to the health portfolio, and felt it prudent, with my agreement, for the leader to offer my portfolio to someone else, and could I recommend a suitable colleague. The first person to spring to mind, was my friend in our broom cupboard. It might keep her out of the way once I had past everything on to her. Whilst it was very disappointing, it got us out of a very difficult situation. I wrote back to say that I was sorry if I had let the Party down, but agreed with him it was probably for the best, a least till I get better. From now on the difficulty was, for my husband to be seen, but at the same time keep a low profile, easier said than done. On one of his trips to the commons he bumped into non other than the leader of the party, he asked my husband if I am any better as he had not seen me in the chamber for a while, and past on his best wishes. My poor husband said he didn't know what to say for a moment and just stood there with his brain about to burst, then to his great relief, the leader said, don't forget, wish her well from me, and just carried on his merry way, phew!. I don't know why phew! because we're doing nothing wrong, but phew! anyway. On his trips to London a couple of times a week, he would post the letters from one of the House of Commons post boxes. It's a bit sneaky I know but that's the way we did it. I felt I should do my very best to earn my keep, and to that end I had to push myself to watch as many political programmes as my brain could absorb, and when I felt up to it I would write a few notes about the relevant pieces. My husband would record everything in sight, including all the news programmes and political slots. On several occasions when he was in London, he would record Prime Minister's Question time just in case I couldn't watch it. Sometimes we both recorded the same programmes, but never mind. Still with an eye on the main chance we kept to this charade for quite a few weeks with more and more of my colleagues asking my poor husband where I was, or more to the point, how I was, but the chief whip didn't seem to be bothered, though I'm sure he knew, he just kept sending "the whip" each week. It was always my intention to go to back to Westminster if at all possible, if only a couple of times, then it came back to me. Remember the what ifs, one what if, I couldn't get out of my head, was about the unfortunate quirk just came back to haunt me, remember the one, win the seat, but be not well enough to travel down to Westminster to swear the oath and sign the test roll, shit!. When your so ill with cancer as I am becoming with each passing week, a week is a long time, not only in politics, but for me. The timing is even more critical. I would hate to be so near, yet so far, yes, your ahead of me aren't you, two hundred thousand pounds missed by a whisker!, a r r r r r gh!

The date of the general election was back on the agenda once more. Anytime next year will do fine as far as I'm concerned, callous as it may seem I will be long gone by then, and I'm sure I would die happy in the knowledge that my husband and two sons are well placed thanks to my gratuity, and if I was able to stay well enough to retain my seat, that would be a real bonus for me on a personal note. But if I had to stand down, and loose all that money, which is the very thing I don't want to do as I said earlier on in my account of things. Yes I know it sounds bad but we have paid our taxes all our working lives, so what's the problem? and in any case, we're doing nothing illegal, I don't have to sign anything to say I'm fit and well, or tot up my attendances for my constituents to see, or prove anything to anyone. My husband continued to go down to Westminster a couple of times a week and managed to fend off any difficult questions about me, or my state of health, many of them knew that I had had a double mastectomy some years ago, and maybe some of them suspected that it may have returned, particularly my female colleagues with whom I discussed cancer issues many many times as my party's spokesperson on health. After wondering about our situation one evening, we both realised at about the same moment, that most people were wearing reverential gags, not just my colleague with whom I share an office, but as I said, you wouldn't dare ask a nun if she is a virgin. Keep this up and we could be laughing all the way to the bank, well, not me exactly, but you know what I mean. This is almost a roll reversal of the early days when my husband was the one left at home and I was the one swanning off to London, but I didn't say anything to him as things were difficult enough. He put on such a brave face, and just got on with the task in hand. I say task for that is exactly as we saw it now. Frustrating as it was, after all I had worked for and believed in, both on the local council and in Westminster, working for the good of others. Now it's pay back time. There were times when I thought that I really ought to pack it all in, but how, and when. Apart from anything else it would be expensive. It wasn't an option we wanted to even think about. We just had to tough it out for as along as we could. By now we were in a very odd position, we had to keep up a pretence of normally at home as well as Westminster, I had to take my surgery's when I felt o.k. and more to the point, looked o.k. The last thing we wanted, was someone asking questions about my health, specially if an election is to be called soon, and we were not prepared. There were more and more occasions when one of my council colleagues, or my husband had to stand in for me to take my weekly surgeries. Some of the excuses were quite laughable, and at the same time, rather ironic to say the least, on one occasion my husband told a very impatient man that I was writing a paper on cancer treatment and the devastating effect on the family unit, and how they could play their part within such a difficult situation. By the time my husband had given the reason for not being there, he was past caring and muttered something about bloody politicians being a waste of time, and left. Well he can't please all of the people all of the time, but he's doing his very best. Another time he said that I had been up all night and still felt sick, little did they know the real reason for my being sick was the session of radiotherapy I had yesterday, and I was feeling a bit under the weather. Another time he told a very irate sixty something to stop picking, and moaning on about next door's children, as life's to short. See what I mean about the irony. We managed to keep up this deception, for that's exactly what was, a deception, a means to an end. An end that was for one reason, and one season only, the desire to leave the loves of my life well provided for. I have absolutely no qualms about what I am doing for the benefit of my family, as any mother would, but in particular for my two dear sons, by now, grown up young men, both destined for a bright future. I do hope the oldest, by twenty minutes will enter politics later in life as he has intimated on more than one occasion, I am sure they'll do well whatever they choose to do. Then I had a horrible thought, I do hope my actions over the next few weeks and months won't jeopardise any ambition they may have in years to come.

RED CARPET

The man from the Garden Centre rang on Friday to say our tree had been delivered to their centre. My husband explained the situation, and they were very obliging. I can't wait to see it, but even better than that, my husband had arranged for boys to come up for the weekend. On a nice sunny Sunday afternoon in late February, we had the tree planting ceremony. At two o'clock a pickup truck arrived with my tree, two very nice men from the garden centre came along to plant it for us. My husband had found a piece of red stair carpet in the loft. With the boys and their dad looking so proud, so happy, and loving me, in a way I never thought possible, we were all fighting hard to hold back the tears, he said the nicest thing, he said as he comes down the road in the spring he will be able to see the blossom, and think of me. Then through a vale of tears he said, but you will always be in my heart even when the blossom falls. After the men from the garden centre tidied up and left, the boys rolled out the red carpet for me and dad to stand on as the groud is a bit sodden at this time of the year. We had a toast with champagne and a few minutes reflection. As we were turning to go round to the back garden, I don't know why, or even how, I could say or even think such a thing. But for some reason I said to my husband that time seemed to be dragging, the moment I said it, I was horrified to think what I just said to my poor husband. I'm sure he was on the verge of tears, but he managed to go to the end of the garden and pretended to carry on with what he was doing, I felt so ashamed I went into the house, and left him alone for the while. I really knocked the wind out of his sails. I'm glad the boys didn't hear me. Sunday evening was not as I would have liked, but I suppose that's my fault. The boys said did I mind if they went for a drink, not at all, and take your dad too. I was glad of an hour or so of quietude. The next day, Monday, all three shared a taxi to the station, as they waved goodbye, I think he was glad that he was due to go down to London. I managed to convince him to stay for a couple of days by telling him I was feeling o.k. and could look after myself perfectly well. A couple of days turned into four. Don't get me wrong we didn't have a falling out or anything, it was a mutual feeling that the situation was beginning to get on our nerves, and a few days of reflection would do no harm at all, between you and me I think he was getting used to the idea that I wasn't going to be around for to much longer. He rang me morning and evenings just to say what I already knew, that he loved me just as much as on our wedding day. On Thursday morning he rang to say he was catching the one o'clock train and would be home about three thirty and have the kettle on because he had some news for me. From the time of that morning 'phone-call till three thirty, seemed a lot longer than the six or so hours that it actually was. The good news he mentioned, could only be about the date of the next election as there had been a lot of speculation in the press, not to mention the whisperings in the tea rooms and bars in Westminster. Just before he was due home, on what was a nice spring evening I managed to get myself to the end of the drive to greet him. When his taxi arrived after only a few minutes wait he was not best pleased to see me standing by the gate, by now looking sad and in need of a coat of paint, not me! the gate. He didn't say anything but I could tell he was a bit concerned that someone might see me. Not that I was in hiding, it's just that my husband didn't want us, or more to the point, me, to be caught off guard. Oh! your waiting to hear the news, as we sat at the kitchen table with our obligatory cup of tea, he said there's a strong rumour, and one that is gathering momentum, that there could be a spring election, that's great I said, can't wait. We want a election soon, whilst I'm still fit enough to stand for re-election. Alternatively next year for obvious reasons. The question now is, will I be well enough to convince the party selection committee when the time comes to put myself forward as their candidate? Candidates are usually in place long before an election is due, but as I am their sitting M.P. I am naturally expected to stand again. They knew I had not been too well for the past few months, and some of them knew I'd had a double mastectomy some years ago, but just how much, or how little or even how many knew, was difficult to find out, short of asking someone outright, only time will tell. It'll soon be March, and with Easter fast approaching meant a release from all the deceit and pretence for a week or so. In the meantime my husband continued with his twice weekly visits to Westminster. On his return home as I watched him walk up the drive I could tell by his demeanour if there was good news, or not. But it just got better and better, with not much action on the front benches, no one asking about my absence, and me feeling quite a lot better than I have in weeks. I really began to believe, that if it came to it, we could pull it off. But not so fast, on one of my visits to the Hospital for what is by now palliative radiotherapy, usually administered by a nurse, It had been a arranged for me to see my oncologist after my treatment. I have all ways asked him to be totally honest with me, now I may be sorry I opened my big mouth. But here we are, back in that little green waiting room, it's looking a bit shabby by now, and needs repainting, but quite frankly my dear I don't give a damn, yes, we've all seen the film. As we waited the usual ten minutes, I was at my wits end with worry. We wondered if we might see our Cavana Q.C friend today, we hadn't seen him for quite some time and hoped he would be here this time. I was just beginning to get ready to run out of the place, well, more like hobble, when my name was called out by a nurse, she checked my appointment card, then took me through see the doctor. I was somewhat disappointed our Cavana QC oncologist wasn't on hand, but never mind, as we went into the consulting room, and sat down in front of the desk, all was quiet for a moment as he turned the pages of my file, by now, we were both getting good at reading facial expressions, and on this occasion the omens were not looking good. He leaned forward on his big beautiful oak desk and gave me little hope for the future. He said he wanted to see me in person so he could make a proper assessment on my situation. He said my weight loss was slightly more than he was expecting at this stage and the condition known as "Chacexia" a muscle waisting condition" was beginning to develop even though my appetite was as good as could be expected. Mind you, he was only going off what I told him. Well, there it was. I had to ask him if he thought that my prognosis might improve, but our hopes were dashed on the rocks once more. He was not as optimistic as I would have liked him to be, he said the original prognosis of last October, was, unfortunately in his opinion, still the case, but you never can be absolutely sure. As we were about leave, I couldn't help but ask where my silver haired Oncologist was today as I hadn't seen him for a while. I wish I hadn't. He said I'm very sorry to tell you that he died of a brain tumour last summer. That news hit me harder than my own news. I couldn't believe that Him or Her up there could be so cruel. I cried all the way home, not for me or my husband, but the injustice, that such a dedicated doctor could have his life cut short in that way. When we got home, my husband told me he had read an article in the paper last August about our oncologist. He said he didn't want me to see it at that particular time, as I was in the middle of my second battle with this visitation that had seen fit to tap on my shoulder once more. As you can imagine the boys are very worried about me. Their father rings them every two or three days, just to let the know how I 'm getting along. He thinks I don't know it, but he rings them when he thinks I'm asleep, bless him. It's now mid March, if my doctor is right with my prognosis, for the timing in these matters is by no means an exact science, gave me about ten to twelve weeks to live, I can't believe I just said that!, but it's something we've have got used to by now. If this election follows the same four year term as previously, as it might, then a May election is on the cards, that gives us only a few weeks to fine tune our plans, a plan to cover every eventuality. One eventuality will be, how soon will my weight loss begin to be noticed, it seems funny now because when I was pregnant with the twins I was wondering how soon my weight gain might be noticed. Back to business, as I was, by now, beginning to feel to scared to take a chance to go to Westminster, as I might take a turn for the worse, and make things even more difficult, and finnish up having to admit to the party bosses hear and in Westminster, that I was not going to beat this cancer, far from it! I was on a very fast train to nowhere. So it was up to my husband to keep up the pretence of normality in all things political. However on one occasion, the Party Leader did say to my husband that he hoped to see me in the "House" soon after the election, and to tell me to keep my pecker up, as the election could be sooner than people think. In one respect this was a piece of news we didn't quite know how to take. The trick is to get myself adopted as the candidate for my constituency whilst I am still looking reasonably healthy. We couldn't leave it to long in case my health got to such a point that it would be impossible to convince the local party members that I was well enough to run for election. What we didn't want, was someone asking me about my health. Of course it would be unlikely to happen, thank God for political correctness. If people did start to ask questions we could wave goodbye to over two hundred thousand pounds if I am forced to step down, but be in no doubt I'm determined to go through with this plan right to the bitter end if there is a chance we can do it. The odd thing is, there are no barriers to a candidate wishing to run for election, even if they are indeed very ill, and here's another oddity, you don't even have to canvass, so long as the party officers go along with it. A candidate could apply to stand in a constituency, then, as long as they do the legal bits they can fly of to the Bahamas for the duration, and no one can say a word. If they win, all that is required of them, is to take the oath, sign the test roll, and collect their salary. Nice work if you can get it. And you can.

A CUNNING PLAN

One Sunday afternoon, we devised a cunning plan. We needed to jolt the constituency party chairman into action. A deception by way of my husband telling him he had good information that the general election was to be called sooner rather than later, even if he were wrong it really wouldn't matter to anyone would it. They think it could well be next year. This deception was aimed at getting our constituency chairman to call an executive committee meeting as soon as the rules will allow, hopefully for sometime next week. The hospital gave me pain relief in the form of morphine tablets, and they were a God send. Still no definitive news about the election. The good news is that we had a calling letter and the minutes of the last executive meeting along with the agenda to attend an executive meeting on Friday evening of next week. I just hope I'm feeling better than I am right now. On Sunday afternoon I felt awful, my back was racked with pain, my legs felt as if they were not a part of me. I stood there in the bathroom and looked in the mirror at that strange emaciated woman looking back at me as if it were some one else. I was so wretched I took two tablets with a glass of water instead of my usual one. I knew two tablets wouldn't do any real harm but I was curious as to what the effect might be, so I went for a lie down with my husbands blue dressing gown on, it's getting a bit moth eaten by now, but I love it. The next thing I knew was the sound of my husband running up stairs two at a time, I thought the house was on fire, the fear his voice was clearly palpable. The reason for his concern was the fact that there were no lights on in the lounge, as he would expect, at six thirty in mid march. It seems I had been fast asleep all afternoon. As I gathered my thoughts and slung my legs round and put my feet on the bedside rug, I had a feeling that I am sure people get after visiting, Knock, or Lourdes, a feeling of elation and surprise and complete, well, elation and surprise! I felt great, well, perhaps great is to strong a word but I did feel a lot better. I'm not sure if it was the extra morphine, or the good rest I had. I think maybe a combination of the two would be a more likely reason. My husband couldn't believe what he saw as I walked across the bedroom and gave him a big hug. I could tell he was pleased, but when I told him about the extra morphine dose, he asked me to promise him, with all my heart never to even think of such a move again. I could tell he was fearful of what might have been. The next day, Monday, I still felt good, and wondered, should I 'phone my oncologist, or just say nothing, I decided on the latter. It was only five days before we had to meet all the members of the executive committee at close quarters for, hopefully a "rubber stamp" adoption of me as their candidate for the General Election. I sent a detailed account of my work as their Member of Parliament. Detailing as many instances as the data protection act would allow about constituents I had helped and advised over the years as their M.P. and how I wanted to serve for another four to five years, oh!

and chuck in a couple of marathons for luck. We nearly made a very serious tactical error, I did wonder if I should mention my voting records, to emphasise what I had voted on whilst serving as their Member of Parliament, and then realised, that if anyone looked it up an the web, it would show that for most 2005 I'd been absent from the vote, and would look as though I had been absent from Westminster all together. Truth be known, I was absent, but I don't want anyone looking to close do I. Even though it is not an official website, people could get the wrong message.

JOHN F. KENNEDY

Friday came all too soon, we hoped that we had prepared our strategy in the event of any awkward questions, and tried to cover every eventuality, but we couldn't be sure how long the meeting would be, or if anything would go wrong. We left the house as late as we dared, on what was, a horrible wet blustery evening. Before we go any further, I'm going to let you into a secret, it won't take a minute. On cold days, the late President John F. Kennedy could often be seen wearing only a light suit, and often no overcoat or top coat of any kind. This was to give the impression that he was fit and healthy, but what people didn't know was that he was wearing long thermal underwear to keep warm. I'm not going into to much detail, but suffice to say I was well prepared. All I needed was a wool frock to give me, I never thought I'd say this, a bit of bulk, and a lightweight top coat, so far so good. It was, in many ways, make or break time to see if we could convince the executive that I was well enough to carry on and fight an election campaign and what's more retain the seat. If the subject of my health was mentioned, even though it shouldn't be, I'd say I was getting better, well, not getting better, best not say that, just say I was feeling better, semantics, I know, and say how I am looking forward to the Election, whenever it comes. Short of telling barefaced lies, there wasn't much we could say, just put the best possible spin on the situation, and say thank you, and hope the morphine would do the trick. As we pulled up by the door of the club the chairman came to greet us as he hung onto his umbrella like a man possessed, and ushered us in saying that there was a poor turnout on such a terrible night, good! I thought, as we went inside. There were a few hello's and friendly nods of approval as we went into the "lion's den" and sat at the far end of the concert room with what seemed like the Spanish Inquisition before us. We sat at the designated table to one side of this typical bad taste, nicotine stained club room that has been the butt of numerous jokes, over the years, no pun intended. At eight o'clock sharp the meeting was called to order by the chairman, whilst the secretary took the minutes, the chairman said he his usual opening remarks and thanked everyone for there attendance, all thirteen, on such a wet and windy night, "get on with it I thought". After everyone had signed the attendance book, and read the minutes of the last executive meeting, we all agreed that the minutes were a true record of the last executive, with a show of hands, he moved to the next item on the agenda, to our surprise he said as there's only one item on the agenda, as the others had been moved forward to the A.G.M. I was delighted, because the agenda we received gave four items to discuss. He said we might as well get on with it, and at that point proposed me to be re'adopted as their Parliamentary candidate, and therefore to stand at the next General Election for the constituency, he then called for a seconder, and to my surprise several hands went up, all in favour, all hands went up, so the motion was carried unanimously at eight thirty, wham bham thank you mam! It couldn't have gone better in my wildest dreams. Well, not so fast, my selection still had to be ratified by the whole constituency party membership, even though the chairman said that it was a mere formality. I still have it to do. I was now expected to say a few words, can you imagine, me! a few words, well on this occasion, least said the better. When we arrived at the club, I started a diplomatic cough just in case I found that I was not going to be able to speak for to long and had to ask my husband to use my notes and speak for me. However, as the meeting was very short and I felt fine. I got to my feet, the whole room seemed to look at me as one, or thirteen, and hung on my every word with renewed enthusiasm, and expectation. I think I gave a good account of myself, and whats more, I loved every minute of it, till my husband, said that we should not push our luck, for that was exactly what we had that evening, in spades. One thing we didn't anticipate was that the meeting would be so short, done and dusted well before nine o'clock, good in one way, but not in another, a few of them knew I'd had cancer some years ago, so we didn't want to hang around waiting for someone to ask any difficult questions about my health. Then, ever the diplomat, I put a fifty pound note on the bar to get drinks for everyone, and said my husband had some paper work to do, with the fifty pound note as a distraction, we made our excuses and left with a huge sigh of relief that I felt could be heard a mile off as we sped from the club car park on that wet, but magical night in March 2005. When we got home the realisation of what we have just pulled off, was just incredible, so my husband nipped out for a chicken curry and fried rice to celebrate. A few days later,we had a calling letter to attend the meeting of the whole membership of the constituency party members at the A.G.M. a week next Friday to ratify my selection as candidate. Remember the chairman saying it would be just a formality. If I'm not too well, it could easily end in disaster.

The meeting is just over a week away and I had no way of knowing how well I may, or may not be. The chairman rang during the week, and as luck had it, I was upstairs having a lie down, so my husband answered. Being a nosey parker I stood on the landing trying to here what was going on, but couldn't make sense of the one sided conversation, I didn't have to wait long as I heard my husband say thanks for calling and put the 'phone down. I called down to him to see who it was, he said hang on a minute I'll bring you a cup of tea. "Bloody hell" I thought, whats wrong now, over the past few months bad news has always been preceded with a cup of tea. When he came into the bedroom he didn't look best pleased. As he sat on the edge of the bed, he said it looks as if all our efforts may have been for nothing. NOTHING! what do you mean, NOTHING? I'm ashamed to say I really lost my rag with him. Please tell me you're kidding. In the few minutes since he put the 'phone down and brewed up, to this moment as he sat on the bed he had had time to think. Well maybe it's not that bad. Will you please tell me who was on the 'phone, the Constituency Chairman, he said with a real sense of foreboding. "Shit"! what did he want?. Do you remember at your adoption meeting? he said your adoption by the whole constituency party membership would be just a formality, keep talking

I said, with increasing trepidation. Well it's not quite as simple as that, though it's not in the rules, you are expected to attend the constituency A.G.M. and you'll be expected to address the meeting after your formal adoption as candidate by the full membership as it says in the agenda, what did you say? I told him you were out visiting a friend, and I would get back to him. Did you say the meeting's a week on Friday, I think that's the day after my 'therapy, you know I'll feel rotten! What do we do now. My husband was due to go down to Westminster in the morning, I said, sod Westminster, this is far more important. It's a good job one of us had our thinking cap on as he said, it's just as important we keep to the plan, so I'll go to Westminster as usual, grab the post, then I'll get an up-to-date copy of the party's rule book, from the whips office, I'll say it's for a constituent. Then I'll nip to the flat for the junk mail, have quick look round, and catch an early train home. During which time I'll read through the latest set of rules, so we can compare the rules we have here, to see if they differ in any way. After the initial shock of what could be a disaster, I thought, sounds good to me, so we had a nice cup of tea. A little later we watched one of our all time favourite musicals, "Carousel" with Gordon MacRae and Shirley Jones. Even when I was fit and well, I would cry buckets, but now, for some strange reason I felt more able to deal with the tragedy as struck poor Julie when Billy Bigalow was stabbed to death. Sounds morbid, I know. The film begins with Billy Bigalow polishing the stars and looking down on Julie and the young daughter he's never seen. I could just imagine myself polishing the three biggest and brightest stars in the sky for all eternity. One for my ever loving husband, and two identical stars for the twins. We went to bed feeling a lot more relaxed than I could have thought only a few hours ago. The next morning we set our plan in motion, with hubby off to Westminster, me feeling a lot better, must be the Adrenaline, or something in the air. Something the air! that's it, "Spring" My Malus Domestica was just beginning to show the world what she could do, with lovely white, pink edged buds just beginning to come through. The only way I could think of to pass the time would be to cook a nice tea for him. At four thirty on the dot I heard the taxi door slam shut, as I looked through the net curtains, I could see by his jaunty step that we were most likely on the same wave length. The minute he came through the door he said, smells good, pity I'm not hungry, then with his big beaming smile gave me a big hug, and said I'm bloomin' staving. As we sat down to roast beef and the usual veg' we couldn't help ourselves but to ask one another what we found in the rules. Right, I said, me first, the chairman said it's not in the rules that I should attend a full meeting to ratify my candidature, but it would look bad if I didn't, so we'll have to be a bit hard faced here, and box cleaver. After reading though the rules several times, we hit upon the paragraph that says, "a sitting Member of Parliament shall be required to make written application to the Executive Council should He or She wish to seek re-adoption to stand again for parliament. The Member of Parliament shall be kept fully informed of the association's intentions and shall have the right to be heard by a general meeting of the association before any steps are taken to consider any other candidates". That's the bit I like, any other candidates, there aren't any!. We swapped rule books and read through each in turn just to be sure we were on the same wavelenth, and we were. That's it then, we're going back to stage one. I'll make a written application to the Executive Council through the chairman as it says in the rules, and say that as I understand that there are no other candidates up for consideration I feel that this application is valid as per rules of the association. We thought it best not to say weather we were going to the meeting, just in case I was not up to it. Just fancy, there was no need to go to the meeting last Friday after all. But maybe It's as well we did. We managed to come up with my own glowing account of my previous four years in Parliament, and my desire to carry on and serve the constituency that I love so much, "sounds good ah" I went on to blow my own trumpet long and loud for two pages of A4. When we were both happy with it, we folded it nice and neat, and put it in envelope, we each had a lick of the envelope for luck, we needed it, as I had no intention of going to the jaws of the A.G.M. My husband posted it through the constituency office door by hand, just to make sure there no slip ups, as my adoption would be automatic once an election is called. I'm beginning to loose track of time now, but I think it was Tuesday, as my husband opened the post as usual, he could see one from the association office, as he read it his eyes lit up as he breathed a huge sigh of relief as he said, "were in", What do you mean "were in" you're now the adopted candidate for this constituency!. What a relief, I think I'll say it again, what a relief, there was no way I could have gone to that A.G.M. as expected, I didn't feel like going to the meeting after my treatment. This meant we could put that particular problem to bed for a while. You may wonder why I was so preoccupied with getting myself adopted as the candidate for the election. Well, It's just one more obstacle out of the way. When the election is called, my adoption as candidate kicks in automatically, and the activities of the Association and its Branches are suspended until after polling day. Of course we can still use the constituency office to run the campaign for the duration of the election, and nothing else. That's why we have a separate bank account for the election expenses. The officers for the election are appointed by the election campaign team when the time comes. My agent is usually chosen by me, as we have to work closely together, as he or she, is in effect is my boss during the campaign. This person could be me, or my husband, but, as I may not be too well during the election, and my husband may be to busy covering for me, I will chose someone very carefully. The treasurer is usually appointed by the executive to keep an eye on the money. It's usually the incumbent constituency treasurer.

As Easter was fast approaching, and following the unwritten rule of no electioneering during the Easter period, particularly on Good Friday and Easter Sunday, we could look forward to a little restbite, not so much for me, but for my husband with his too-ing and fro-ing to Westminster two or three times a week to keep things ticking over. What a star he was, especially as I was beginning to feel worse week by week. It seemed to come on so fast, so fast in fact that I made an appointment to see my oncologist before my scheduled appointment in a few weeks time, just to get a better understanding of my prognosis. Yes you get the picture, that little green waiting room, well, you know the rest. My oncologist said his original prognosis regarding my cancer was as expected at this stage of my illness, and that he was sorry he did not have anything more positive to offer, and we should prepare for the worst. One thing I was glad of, my husband didn't see the doctor writing in the margin of my file, what looked like, "end stage". but I couldn't be sure, you know what a doctors writing is like, and I certainly wasn't going to ask. As there are differing methods of describing the stages of cancer, be it first, second, and so on, the fact remains that things were not looking good, we could only guess when the election would be. It could co-inside with my illness becoming more apparent, and I would have to stand down as candidate and loose two hundred thousand smackers. When we were at the hospital I told the doctor about the time I took two painkillers. He was silent for a moment, and then said that all he could do was to warn me of the dangers of over doing it, but he was resigned to the fact that it was me that was suffering and he could not in all consciousness tell me not to do it. Cancer is about the worst illness anyone can have, in the beginning you think why me?, why not some paedophile, or murderer, it seems so unfair! Then you begin to think, I'll beat this thing, this monster, this visitation. But I soon began to get used to the idea of dying, and was more and more determined to make the best of the hand I've been dealt. I have no way of knowing, but I suppose if I was in my eighties or nineties it would be easier to cope with, but I don't know. Today I feel so wretched, tomorrow I may feel like going to the shops, or even Westminster, ah, Westminster, it seems ages since the "Intake dinner" last December. Last time I was in the commons I nearly feinted, remember that? I am so bloody frustrated. There's so much I want to do, I want to bring greater awareness of the devastating trauma a woman suffers after a mastectomy, and fight for those less fortunate than me, with my good salary and even better expenses package, not to mention free first class train travel, and more free stationary you can shake a stick at. Sorry I'm starting to witter on a bit. The weight began to drop off me at an alarming rate of about two and a half pounds a week and I began to think the unthinkable may well happen, and ruin everything if, as we expect, the election is called for early May. Don't tell my husband, but I thought that if I was to die sooner, rather that later, then things would look after themselves, but when I thought about it, I soon realised I was too much of coward to take an overdose, besides, my husband would be devastated and I wouldn't wish that on him or my two boys, then I realised that he could well miss out financially as suicide could well negate my Parliamentary gratuity for my husband, not to mention my own life insurance. So I put that idea to the back of my mind "not one of my better ideas" and a good job too, because the next day I felt good again. When my husband was in London I did my best to keep busy by writing pieces for my election literature, and to sort out a few photographs for my campaign leaflets. I dare not and could not use any resent pictures of me, not that there were many, just half a dozen from Christmas and the tree planting ceremony, as I was decidedly slimmer by this time and looked, well to say the least, not in the rudest of health. The only other photographs I had were of me during the last election, there was no way I could use them as it's possible that someone may object, or worse, they could be no longer with us. I suppose we're all the same when we start to look at old photo's. Lots and lots of the boys growing up, and some black and white ones of me and my husband on the beach at Grange over Sands, my first election campaign and newspaper cuttings of the foot and mouth outbreak and my first trip to Westminster in 2001. After about two hours had flown by, I managed to sort out half a dozen that were not too ancient and put them to one side. When I realised what time it was I thought I'd better 'phone my husband, he didn't 'phone me now, for fear of disturbing me in case I was sleeping, just to tell him what he already knew, I love him, yes I know it sounds a bit soppy but I'm glad I did ring him when I did, it was the best news he could have given me. Well I think it was, he said that the date of the election was almost certain be early May. The prime minister is due his usual audience with the Queen next Tuesday and is expected to request disillusionment of Parliament and call a general election. The poll to take place in May what a relief that was. Of course this information was given to my husband in strict confidence by our party leader, don't say anything but my husband said he thought that he was a bit tipsy, but he did have friends in high places, so there was no reason to doubt him. The worst of it was, we couldn't say a word to anyone, we couldn't do any more than the usual sorting out those people that wanted postal votes, this was only "piece meal" as we didn't have the latest register of electors. We did manage to organise some cars to be available on polling day, for lifts to the polling station, and try to be ready for the off! My husband woke me with my usual cup of tea at about eight o'clock on the Tuesday morning, but didn't stay with me to long, he was glued to the tele' watching the BBC news for just a hint of the Prime Minister's intentions for the day. Speculation was the order of the day, or so it seemed. Then the news programme switched to Downing Street, and to our tremendous delight the Prime Minister announced that he was going to Buckingham Palace to ask the Queen for permission to dissolve of Parliament, and call a general election, but did not say when. But we were sure it would be the 5th of May. With an enormous sigh of relief we hugged each other and cried, well, I did. This is where my husband had to be very careful what he said about my health, or how I was feeling. We did have a very awkward few minutes the day after the election was called. My newly appointed agent called at the house to see me and my husband, luckily I was upstairs having a lie down when he called, come in, my husband said with a smile in his voice, hiding the fact that he was thinking on his feet like no one had ever done before or since. I was, by this time trying to puff up my cheeks slap a bit of lipstick on, just in case he needed to see me in person, but I couldn't think why, then I remembered, I was required to sign my nomination papers and other papers for his appointment as my agent. Who, by the way, was a very nice man and the soul of discretion and would turn out to be a good choice. I could have asked my husband to bring them upstairs for me to sign, but thought it would be a good move for him to see me looking reasonably well as I am at the moment. I had my husbands big blue dressing gown on, yes I know it's getting a bit tatty now but it gave me a bit of bulk, I wet the front of my hair then put a towel over my head to make it look as if I was washing my hair, talk about thinking on your feet, getting good ah! I duly signed the papers and ask him, as he was so enthusiastic, to get the rest of the nominees to sign for me as I was very busy with some constituecy work. A request that seemed to put a feather in his cap, a spring in his step, and a sigh of relief from the both of us, as he wished me well, and good afternoon, all done and dusted in a few minutes. After tea we did wonder if he knew more than he was letting on. Sunday morning was, for some candidates, probably one, of three or four times they go to church throughout the whole year, but I couldn't go even if I wanted to, and besides I couldn't be a hypocrite any more, I think He knows. Canvasing was also out of the question by now, as I would not be able to walk more than a few hundred yards without getting so tired, and how stupid would I be to go canvasing in a wheelchair, even if I had one. It would be foolhardy of me to expose myself to such close scrutiny, in any event my husband told people that I didn't want to expose myself to any infection. Another close call, my newly appointed agent rang me and said he wanted some photographs for the local press as he forgot to ask me other the morning, he wanted to catch Tuesday's deadline, so it's a good job I had some ready for him. My husband met him at the door with the pictures of me, just head and shoulders would have to do for the press, but as for my leaflets I was not sure what we could do. In the event we didn't suggest anything more about pictures, nor did anyone ask. We had a sneaking feeling that those close to me, and had known me for some time, knew more than they were letting on. My agent called on me several times, but to our amazement, never said a word. Nothing about my weight loss, or my inability to go out on the hustings, or see anyone. He even arranged for one of my councillor friends to stand in for me, on the one privilege that I could not believe I had to miss, yes, you've guessed it, a chance to go on the local T.V. politics programme, what a bitter blow after all my work for the local community and all the fund raising for the local hospital and many other good causes in my constituency, I could have gained a thousand votes in ten minutes easily. But if I went on as I am right now I could have lost a damn sight more. My husband wrote most of my letters and campaign articles for my leaflets, if a piece for the local press is sent as an e-mail, who's to know! I did manage to answer the 'phone more often that not, my husband said I sounded alright on the 'phone, as my voice was quite strong, well, that's something, maybe because I had been a teacher for years that my voice was strong, ladies have strong voices, whereas men have loud voices. We dreaded the postman each morning as there was always that feeling in the back of our minds, the possibility that a letter of, shall we say "inquisition" could drop on the mat. After all I hadn't been out and about for weeks, this was something that worried me no end. But none arrived, so far so good. What I did receive, to my great surprise and relief was a hand written letter from the leader of the party saying in no uncertain terms to keep up the good work, and looked forward to welcoming me to the "house" once more when I had retained the seat for our party. We were on edge throughout the whole of the election campaign, specially when my G.P. offered to arrange some nursing help from the local hospice, how would that look if anyone got to know. I had to decline her kind offer and say my husband was at home most of the time. Think on this, my G.P. knew how ill I was, but was bound by patient confidentiality to respect my privacy. The chief whip knew, as he is the "Mother Hen" of the party. The party leader knew, and most, if not all of my party colleagues, must have known, as they couldn't' help but notice I was conspicuous by my absence, and my health port folio had been taken over by my colleague some weeks ago. My husband was running around like a whirling dervish, leaving me to organise where and when I was able. Some days I felt like shit! whoops, I'm beginning to swear a bit to much now, well wouldn't you? The next day I wouldn't feel so bad. We thought it would be for the best if our sons didn't come in direct contact with the general public just in case they let slip the "C" word to the wrong people or worse still, the opposition candidates. I'm damn sure they must know too. So we kept their activities to running errands, putting leaflets in bunches with all the streets named and made ready for the deliverers. As far as anyone knew our sons were busy, and had their own lives to lead.To our amazement and relief, no one said a word about my illness or how I was getting on, just to report that things were looking positive and that I could inrease my share of the vote.

There seemed to be an underlying feeling that the electorate were sympathetic towards me as it was common knowledge that I had had a double mastectomy some years ago, but what they didn't know, was that I have been fighting a second battle with cancer these past eight months or more, nor do they know just how serious it is, and my prognosis is very poor indeed, well,as only we know it's terminal. I'm sure that most of my deliverers knew that I am to ill to deliver, or campaign in person, but as I mentioned earlier, I think they were wearing reverential gags, or blinkers. I wonder what they would think if they realised we were spending their hard won campaign funds so we can collect a quarter of a million smackers, tax free. Remember when we were in Westminster, it was the same thing there, with all the self imposed gags and blinkers. Like asking a nun, well, you know the rest, a position that suited us no end. On the Sunday of the second week we had a very tricky time. Because of my non appearance on the Sunday lunchtime politics programme. About two o'clock, the 'phone rang, trying to think who it could be, answered it, to my horror it was the local news paper's political editor, asking about my state of health. He said he watched the programme and was surprised that I had appointed a councillor to stand in for me. He said he 'phoned the programme editor and was told you sighted health reasons for your absence. I was "gob smacked", he wanted to know why I didn't appear on the lunchtime politics programme, especially as I was the one going on about the lack of exposure for newcomers back in 2001. Serves me right I suppose. He also asked me if I was expecting any "top brass" from the Party, such as our leader for instance, or maybe a front bencher. After all, I did have a very small majority. By now, I have become very good at thinking on my feet and said I didn't consider myself a newcomer anymore, but didn't want to let the side down, so I asked my councillor colleague to be our cheerleader for the party in this area as a whole, as there's more than one constituency being contested by my party's candidates in this election, and contrary to popular belief we politicians like to be fair. He continued to push me about my health, and the lack of celebrities, not seeing me out and about as much as they would expect in such marginal seat as mine. This was indeed a very awkward few minutes. He knew I'd had a double mastectomy some years ago because his paper gave me a lot of column inches during the local council elections, and my fund raising marathons long before I became an M.P. I was a bit of a curiosity at the time because I had the cheek to tell my opponents that I knew I was going to win hands down, big headed I know, but it worked for me at the time. Now it's come back to bite me on the bum. I managed to convince him that I was well enough to run for re-election and would be taking my seat as soon as I was fit enough to do so. I was just about put the 'phone down, when I had a great idea. I managed to fob him off with a promise of a statement regarding my health. As that seemed to pacify him, he thanked me for my time and said he looked forward to my statement, and promised to print it verbatim. As the 'phone fell silent, I wasn't sure if he'd hung up or not, was he trying to catch me out, was he listening for an unguarded word, or a sigh of relief. I quietly replaced the receiver and sat down for a few minutes. When my husband came home after taking some leaflets to one of our deliverers I told him about the political editor from the local paper ringing me, as I explained what I said to him, he seemed to think I did the right thing, it's to late now anyway. After tea we put our heads together to hatch a cunning plan, we had to come up with a form of words that would satisfy the editor, and not least, the voters. The most important thing was, how it would appear on the printed page, would it be a separate panel, or highlighted in some way, then, at precisely the same moment, we both said, "lets bury it". By bury it, I mean loose it in the middle of a political diatribe on the current political scene, and my election hopes for the future, then slip in my "health statement", right in the middle. Sunday afternoon was the one time we could almost guarantee a bit of piece and quiet. I was sad to see my Malus Domestica, remember, the apple tree, looking a bit dishevelled as the blossom began to fall. It may seem a bit silly to you, but I said to my husband, I'm going to have bath now, and what I would like you to do for me while I'm upstairs, is to brush up any bits of blossom off the lawn and maybe trim one or two of the branches that have gone a bit astray, and make it look as though we care for it. I had a lovely warm soak in apple blossom bubble bath till it began to cool a little. I managed to get out of the bath on my own, much to my husband's consternation, I finished drying my hair and gave it a gentle brush, I want to hang on to as much as I can these days. As I looked out of the bedroom window I just caught a glimpse of my husband going round to the back garden with a bag of rubbish, then he called up to me to see if I was o.k. getting out of the bath. I said I'm nearly dressed and would soon be down to make a brew, so put the kettle on. I cannot repeat what he said, it was a bit naughty. I felt better, the tree looked even better, and my husband look pleased with both of us. Even though it was mid April it's still a bit chilly in the evenings, so I kept the big blue dressing gown on as we had our cup of tea and a think about my health statement for the newspaper before Tuesday's deadline, but we didn't need write anything just yet. After tea we watched a bit of tele' and went to bed about ten thirty. Monday morning was nice and peaceful, after breakfast, as it was a good drying day, we stripped the bed, and along with some underwear and few shirts, I loaded the washer then had another cup of tea. Speaking of underwear, I had to ask my husband to get me half a dozen pairs of smaller knickers from Marks and Spencer's. I thought he might be a bit embarrassed, but he said he had no problem getting them for me. As we drank our tea, we had a think about how we were going to formulate my health statement for the press. I could say, I have a good constituency team round me, and a number of friends that are willing to work along side me and my husband to help retain the seat for our party, chuck in a few hopes an aspirations, and that I was feeling well, and raring to go. I suspected it would not be enough for the editor, he may want me to expand a little more on my health, and maybe my cancer, but I was far to experienced to fall in such a trap, that's why I emailed it just an hour before the deadline so he didn't have time to get back to me, if he did, I would be out visiting someone or something. We felt quite confident that our little plan would work. One of many little plans I might add. The election campaign proper is set to run for about twenty three days, I remember previous elections with a great deal of affection, and longing for some strength in my feeble body. What makes matters worse is the fact that on good days my brain is as active as ever, but on bad days I feel so wretched I could just curl up and die. It's so frustrating not being able to get out there and get at 'em!. But who am I kidding, I wouldn't last five minutes out there. My husband, bless him, hired a wheelchair from a medical aids hire shop. We didn't want all and sundry to know how ill I am, so he went in the afternoon and got back before any neighbours came home from work, and whisked the chair through the front door pretty damn quick. Without thinking I asked him how long had he hired it for, I seemed to catch him off guard. He hesitated and said it was an open ended agreement. As I sat in that wheelchair, it really struck home just how much weaker I'm becoming as the weeks role by. The chair was a God send, it was the type with two big wheels, and two small ones at the back, self propelled, I think it's called. It was so light it meant I could get around the house, or sit on the patio and soak up some sunshine while I still had the chance. I found it easy to drive, if that's the right term.

GOODBY OLD FRIEND

I could no longer drive my faithful little mini any more. A fact that saddened me more than I thought it would. Its parked on one side of the drive, with two wheels sinking into the wet soil. It looks so sad with one of the headlights cracked so much so it was half full of dirty water, and one end of the bumper is curved down to one side just a little, it looked for all the world as if it was crying out don't send me away mammy. But it had been there for quite some time, it just had to go. As I watched the man from the local garage with a tow truck one afternoon as he took it away for spares, I couldn't help but shed a tear for my little friend as it was dragged unceremoniously backwards onto the truck, with brakes locked on as if it was trying to hang on to the drive, but it was no use, with a heavy chain thrown over it and made firm, off they went. With my husband's big blue dressing gown wrapped around me, even on warm days I would feel a bit chilly. I would sit on the patio to write a few letters, and listen to my favourite music, mostly Henri Mancini. His music brings back so many happy memories of when we were courting, old fashioned I know, but we really did court each other in those days. I'd listen to the birds, and watch the last vestiges of blossom fall from the trees, and the squirrels running up and down all day long without a care in the world. I asked my husband to get me some nuts so I could feed them, the question was, how long would I be able to feed them. Now then! where was I, oh yes! Do you remember when I said it would be touch and go with the timing of the election, getting myself nominated, running the campaign, hoping that no one would ask awkward questions about my health. Well poling day is only one week away, and the pressure is mounting. With each passing day I feel weaker and weaker, and wonder what the future holds. Not a lot, you may think, you'd be wrong to think that. Well, not wrong, exactly but the future such as it is, is very important to me and my family. A strange quirk of the rules of Parliament come into play soon, inasmuch, that right up to the very time the results are officially declared by the returning officer, my official position is known as a "degraded" member of Parliament, but I still hold entitlement to the death in service gratuity, but once the result is announced, I become a "newly elected" member of Parliament. My salary is paid till the end of the month of the election as of right. But I have no other entitlements including the death in service gratuity until as a new member I have taken the oath and signed the test roll to become an "active" member of Parliament once more. I'm sure I've told you this already, but it's important that you understand the situation regarding the finances. If I'm not an "active member", when I die, we get nowt'. As the day of the poll gets closer, and my spirits were up one day and down in the dumps the next, I didn't know what to do with myself to pass the time, but I'm glad the boys had gone away for few days. There was no point hanging about like a spare wotsit at a wedding. I assured them I would be fine, and told them to go and have a nice break. There were times when I couldn't even get my husband's tea ready, so I busied myself puttin my affairs in order, easier said that done. My husband had to pacify my election agent, and all the helpers, and made sure he offered words of encouragement, but at the same time keep out of the way. It's a good job the weather is dry, a bit cool, but that makes it more pleasant when delivering all those thousands of leaflets day in and day out. My agent was getting a bit worried about me, one afternoon he rang me to express his concerns about my health, as neither he, nor any of my helpers had seen me since the election started, and people were asking questions. Some, thankfully not too many knew I had had cancer a few years ago, they asked him how I was, and did he think I would be o.k. as they hadn't seen much of me recently. Do you remember, when I said we would need a lot of luck. This is just one of those times. I was in particularly good form that afternoon, so I covered the mouth piece of the 'phone and told my husband what the call was about, and should we chance him coming round for a cup of tea, if you think you'll be o.k. then by all means ask him, we could put a few minds at rest. About three thirty I heard my agent crunching up the drive. Then we realised that my wheelchair was in the middle of the lounge, shit! I shouted get rid of that damn chair before he gets to near the window, my husband just managed to get it in the pantry just in time as the door bell rang, my husband let the agent in with a welcome smile to disguise an apprehension you could cut with a knife. I sat on the far side of the kitchen table to hide as much of my slender frame as I could. My top half still looks o.k. with my padded bra. I did my best to reassure him that I was "coping well with the situation" but was under doctors orders to stay indoors for the time being for fear of infection. I'm not sure he was totally happy with my answer, but that was the best we could come up with. After about half an hour of chit chat about how things were going with the election, my husband did his trick of drumming his ring finger on the table as a cue to say we were pushing our luck, so I motioned to my husband to make his move. We've developed a number of exit strategies for times like this. We said we were going out shortly to see some friends, with that he said thanks for the tea, and was on his way in no time at all. I think he was just as glad to leave as we were to see him leave. It's a shame that we have to deceive him in this way, he's such a nice man, but what else could we do. If I confided in him, it would put him in an impossible position. Sometimes it's more difficult to stop doing something you know is wrong, than to carry on. We were on a run-away train and fighting hard not to hit the buffers, the only way out, is the very thing I'm trying to avoid right now, "death". That's the one scenario that would resolve everything. Is that what they mean by catch 22. The cut off day, if a candidate wishes to withdraw their candidature is about ten days before the poll to let the returning officer know so he can put the ballot paper requirements to the printers in plenty of time. I couldn't contemplate such a thing. How would it look at this late stage of my election campaign, Not good I'm sure. The constituency chairman and my agent would be mortified to say the least. All the money spent on my campaign, all the deceit, no pictures, no personal appearances, no top brass, no celebrities.

How embarrassing for everyone, it doesn't bare thinking about. So, as my husband said, keep to the plan. So far so good, things are going to plan, but soon we'll need to cover lies, with bigger lies, ending in a whopper in order to see it through. I must admit, I was a bit worried on Wednesday, because the first edition of the local paper came out in the afternoon. It's up to the editor just how sympathetically or not he writes the piece about me and any comments about my health. Plus the possibility someone had seen me in my chair as I sat in the garden. We live in a modest semi- detached house, overlooked by our neighbours, even through the trees, but they don't bother us to much. But any negative comments at this critical time could have a devastating influence on the outcome of the election. But we needn't have worried, the piece about me was perfect, about two column inches on politics, a few lines about my health, saying I have a good team around me, and how I am looking forward to taking my seat in Parliament once more, then a couple more inches on politics to finish off. In short a nice piece, in quite harmless and very general terms by the political editor, brilliant! It seems I was worrying for nothing, just as my husband had told me all along. I thought he was a bit of a wimp, the editor, not my husband, silly. With only one more leaflet to get out, it was a great relief for me and my family to know that we were almost there. I must admit, I was beginning to feel really guilty about things, with the lies and deception, using people's hard earned cash. Opening letters every morning with cheques in some of them, usually five or ten pounds, but it all adds up. For reasons lost in time, probably something to do with religion, I really don't know, we hadn't used to canvas or deliver on Sundays, but most of us do these days. It's a matter of manpower, as most people can only deliver in the evenings or at the weekends, so we loaded the delivery teams with as many as they need and told them to get in touch with my agent if they had any problems. This day off from the phone calls gave us time to reflect on our situation, and the implications should anything go wrong if people start to ask difficult questions about the state of my health. The worst scenario would be a direct enquiry aimed at me, or my political agent. He's the one in charge of my campaign, and is presumed to be aware of all the facts. So like I said, things were getting a bit desperate. To sum up, I was feeling, and looking awful one day, and the next day I would feel fine, and look fine. I bet if I wore a wool coat and floppy hat, I could get away with a trip out, but I didn't' dare chance it. The pain killers were getting less effective, even increasing the dose didn't help much. Polling day was fast approaching. My next appointment for my treatment was not till a few days after polling day, but I felt as well as could be expected and all was going to plan with my campaign. But I spoke too soon, Monday morning, what a day it was, never in my life have I ever felt so much pain, my back, my legs, all over in fact, my husband called the doctor but he couldn't see me till early after noon. The only positive outlook, strange as it may seem, was, if I died during the next few days, I would receive my gratuity, and that would be that. At one o'clock, precisely the doctor rang the door bell, then pushed open the door and let himself in. With each step as he came upstairs I preyed that he would take away this pain and let me sleep for ever. He sat on the bed and held my hand, after a few minutes he said in an apologetic voice. I'm only a G.P. as you know, but your oncologist has kept me up to date on your situation, and the news is not good I'm afraid. I tried to help him by telling him I know how close the end is He seemed somewhat relieved, as he was finding it difficult to tell me my cancer is to him, no he said quietly as he looked me straight in my eyes as if to ask my forgiveness. I actually felt sorry for him, it must be the worst part of being a doctor. He asked my husband about my medicines regime, then gave me pain killing injection as I found it difficult to swallow my pills at the moment, and sat by the bed for a while. This may sound so callous, but I just had to do it. I pretended to fall asleep, simply because I didn't want him asking anything about the election, or what would the implications be if I die before polling day. Of course he could not say anything because of patient confientiality. I heard my husband ask the doctor just how bad am I, he said I would be lucky to see out the end of the month. Thirty odd days, not long, if I win the election we might just make it. As I said, a damn close run thing. I didn't know what day it was when I woke up, I was out of it. I didn't know if my husband was being extra quiet, as I could hear a pin drop, not a sound from anywhere, very odd! Then I realised it nearly midnight, and only two more days to go before the polls opened, and thank God for that. It will soon be all over, well, this part of the plan at least. Tuesday morning, I never thought it would be quite so easy, even though I was a nervous wreck at times, especially when people called to see me but didn't get the chance. I'll never know how my husband managed to pacify them all so well, and sent them away thinking I was on the mend without him actually saying so. I even got the odd bunch of flowers, and a get well card. It was the same with all the deliverers and poster putter uppers, all wearing reverential gags, and not a word from head office in London, just the letter I told you about, from our leader saying I should carry on if at all possible and retain the seat for the good of the party. By now I thought, sod the party, we're in it for the money. Wednesday, another nice sunny day, we had to make the decision that could make or brake us. Weather or not to go to the count. If we get it wrong, we could undo all we've worked for. With my husband to'ing and fro'ing to London every two or three days, the deceit of posting letters from the Commons, conning the party I'm fit enough to retain my seat. Even though I feel o.k. today, there's no way of knowing how I will feel tomorrow evening. It was a no brainier really. We both decided it would not be a good idea to go to the count, especially as I would almost certainly have to go in my wheelchair, and obviously draw attention to myself. It is still within the power of the returning officer to stop the count, or postpone it, to seek advice. Plus the fact, that someone might express their concerns to the returning officer and create a situation that would be beyond our wildest imaginings and bring shame on my family in buckets. The press would have a field day, my political opponents would remove their reverential gags, they too would have a field day at my expense, and ruin everything we have worked for this past eight months and more. My party hierarchy would not be best pleased to say the least! Not to mention all my colleagues at Westminster, they would all be tared with the same brush. We are only a small party, if a member, especially one with a portfolio on health was absent for months, must be missed. Even though my husband did a sterling job, to'ing and fro'ing every week. So we decided not to go and that was final. Better to be out of the way if anything went wrong and the brown stuff hit the punkah, big time. My agent is there to fight my corner, though what he could do, heaven knows. Polling day, quite a nice day, the weather was on our side once more. My husband got up at just turned six o'clock, he tried not to disturb me as he showers and gets dressed, but he did greet me with a very welcome cup of tea shortly before seven. He had to leave soon, to go to the campaign office and rally the troops as it were. I mooched about the house, worried sick, and waiting for a 'phone-call from my hubby with bad news like, someone asking questions about me, and where was I. Just after one o'clock my husband came home for something to eat, and of course give me a progress report, I felt a bit guilty because I hadn't managed to fix him anything for lunch, we used to say dinner before I became an M.P. I was feeling very sorry for myself and a bit weepy, I had been watching a video of the sound of music, and when the Mother Superior sang "you'll never walk alone" I cried buckets, not for me, but my ever loving husband and our two wonderful sons, it's a good turnout he said. The candidates usually trawl round the polling stations, but obviously I couldn't, and my husband thought if he went round to often people would wonder where I was, so he kept a low profile. Remember, you see one of us, you see us both. That theory was sorely tested, one woman asked where I was, he said I was knocking about somewhere and kept moving. He went on to say he wouldn't be surprised if I increased my majority, well it could hardly be much smaller could it. Any closer would be bad in a way, I don't want to much publicity, but that was out of our hands by now. As it gets later in the day and time to catch people as they get home from work before they get too settled, we begin to organise reminders to our pledge voters that haven't yet voted, by way of a little paper slip through their letter box just to remind them, and phone as many of the rest as we can, and hope for the best. 4. o'clock..... 5.o'clock.....6.o'clock so slow, I know it sounds like I'm wishing my life away but I just want to get this day over with. My poor knackered husband came home just before six, flopped in his big arm chair in front of the tele' to catch up on the news. I managed to make him some sandwiches and a cup of tea, and some for me too. There was nothing much to concern us on the news about the election, just that there seemed to be a good turn out because of the nice weather, but why that should make that much difference, I'm not sure. Someone once thought it was the less well off that didn't have cars and wouldn't venture out if the weather was bad. There would be the usual full coverage on the tele' with the ubiquitous swingometers, and pie charts at the ready, and a look round some of the more interesting seats, not a mention of the marginal seats. Well, you don't need me to tell you what that could mean to me with my tiny majority. If I was to win with a big increase of my majority, the media would be sniffing round in no time, shit! what would we do if that happened. My husband had enough to think about for now without adding to his burden, so I kept quiet. He was just beginning to fall asleep, when suddenly he jumped up and said he was going to the toilet, having a quick wash and out again. He said he would nip home for an hour after the polls close at ten. I wished him luck and waved goodbye, he said he has squared things with my agent and I was not to worry. As he pulled the front door to with a bang I *did* start to worry. What precisely did he mean, squared things with my agent. Did he promise to "see him right" after the election, or did he mean something quite innocent. I decided not to mention it when he came home, or indeed, ever. After a quiet hour or two watching the tele', as expected, I heard his car pull into the drive, the door slam, and his key turn in the front door lock just as news at ten finished. Believe it or not, I actually managed to make him some supper, nothing special, just some spaghetti bolognese, and of course a cup of tea for starters. It took him all of five minutes to wolf it all down, then he sat back in his chair, patted his belly, and said, that was great!. Three words that meant a lot to me, as my one and only aim in the whole of our married life was to please my husband and return his love in every way I can. I think I have done a pretty good job over the last twenty odd years, and though I've not long to live I actually feel happy, sitting here with my husband and no one to disturb us for a couple of hours at least. By now we were sick of all the political meanderings, the what ifs, and who said what to whom as we sat waiting for the 'phone to ring, I thought it seems to be taking ages for the results to come through. We were both just beginning to doze off, when were brought back from the land of nod with the ringing of the 'phone, was it all over, had I won, had I lost, silly really, it's much to soon. As my husband put the 'phone down he played a little trick on me and pretended I'd lost, but only for a second, then smiled and said that was my agent. He rang to say that it was looking good for me and didn't think there would be any need for a re-count like last time. It was fifteen minutes past midnight, a new day, I wondered how many days I had left, enough to get to get me down to Westminster to sign on as it were. I managed to wash up the supper plates while his lordship, through force of habit I suppose, watched the results programme on the tele', with one eye on our constituency. The T.V , people could be at our count, but he didn't speak to anyone that knew anything, he left that to my agent and told him to keep a low profile. I was just about to make yet another cup of tea when the 'phone rang again, I stopped what I was doing, we looked at each other for a moment as it rang out an impatient sort of ring. It half past past one. Before my husband could pick up the 'phone I said a little prayer to myself, I know what I said about God or someone, but I thought I would hedge my bets a little, great! he said as he wrote down the results on our posh portcullis note pad, bloody marvellous, see you tomorrow, thanks for letting us know so soon, and hung up.With the biggest smile ever, well, for a long time, he said just over two thousand votes more than our nearest rival. I said sod the tea, open a bottle of wine. I was speechless for a change as we drank in the election result, and the wine. We went to bed sometime after two o'clock and slept like babies after a good fill of mother's milk. Nothing to do with the wine of course, we were both very tired. Fortunately my husband was up and about early next morning, Friday, it's a good job too, when the telephone rang about nine o'clock, he answered it, but I couldn't quite hear what was being said, so I dragged myself out of bed an put on the big blue dressing gown and went down stairs. My husband turned and looked at me, and said he had a bad feeling about what was about to happen. It seems my agent needed to come round as soon as possible, within the hour in fact, he didn't say anymore, but we had the distinct feeling that something had gone horribly wrong. With that I pulled myself together as best I could, a girl can work wonders with a bit of lippy and blusher, a quick brush of my hair, a wool frock and I didn't look too bad. Ten to ten, we heard the familiar crunch of foot steps coming up the drive, at precisely the same moment we both spotted my wheelchair by the kitchen table. That's twice we've done that, we must be more careful, quick, move the chair I shouted, as my husband grabbed it and pushed it through the back door just as the front door bell rang out like an alarm bell. Come in, what's wrong my husband said wondering what all the panic was about. As he sipped his tea, and calmed down a little, he said one of our counting agents, was asked by a woman, "where's your candidate, and why isn't she here"? Who was she, I don't know. He took hold of my hand, a gesture I could have done without as my hands were by now very frail and lacking grip. I don't know who she was, but your not going to like this, our helper said you were not here because you are in end stage cancer of the spine, shit! I thought, why didn't he tell us last night. Then I thought, maybe not, we couldn't do anything about it. Remember the "what ifs", we thought we had dealt with them all, with my husband going down to Westminster two or three times a week, all the letters written at home and posted in London, the two or more year old pictures in the press, my so called statement with regard to my health. With just one vital piece of our plan to go, is to get down to Westminster before you know what. Then some bastard comes along to spoil everything. This was a "what if" we didn't bargain for, some smart Alick taking an unhealthy interest in my candidature, and whoosh! the brown stuff would hit the punkah good and proper. The problem is, we can't ask anyone, or even say to much to my agent. I'm amazed that he didn't ask me straight out, "is it true"? must be those reverential gags again, or was it the I've seen him o.k. as my husband said earlier, I'm certainly not going to ask him. I thanked my agent, for all his hard work during these past few weeks and was going on to say there's not much we can do now, when he stopped in his tracks. I know what to do, a preemptive strike! That's what we'll do he said, I'll phone the local evening paper, and issue a statement, well not so much a statement, just a few lines saying that though you were not to well this past few weeks, and couldn't go out and about campaigning for fear of infection, and you look forward to getting back to Westminster very soon and thank all the people who voted for you. If I can 'phone in the next hour, I'm sure they'll put it in this evenings issue. Sounds good to me, can I leave it with you then, sure you can he said, and left in a flurry of flying raincoat. As I said earlier, I had an appointment at the hospital on Monday afternoon, for my Radiotherapy, and hoped I would feel a bit better afterwards just in case anyone wanted to see me. No sooner had we regained our composure, when the 'phone rang, who is it now I shouted, the only way to find out, my husband said, is to answer the damn thing! I'm afraid our nerves were beginning to jangle by now, but we needn't have worried, it was a well wisher, just one of dozens throughout the day. My husband answered the 'phone first, and tried to ascertain if they were friend or foe. The former he would hand to me, the latter, well there weren't any. Remember when I was first elected in 2001, I couldn't wait to get to work to see everyone, now I'm trying to avoid everyone. We were just sitting down at the table to have our tea, when the 'phone rang yet again, with a sigh my husband said whoever it is I'll get rid of them. With a few o.k. 's and thanks he put the 'phone down. It was your agent, he said he's shoved the evening paper through the letter box, he could see we were sitting at the table, and didn't want to disturb us, have a look and see what you think, sounds bad to me. On page three there was a picture of me, never thought I'd be on page three, come on now, back to the paper, with my picture looking quite good, a few ambiguous lines saying thankyou to all who voted for me, and that I was looking forward to taking my seat in Westminster very soon to carry on with the job. I thought great! that will do me, short and sweet, we had just about finished our tea when the damn 'phone rang again, who is it now, I said rather angrily, see if you can get rid of them. He said it's the party leader, oh no! has someone complained to him, maybe the woman at the Town Hall last night. Bloody hell! what do I say. He said the best way to handle what could be the end of our deception, was to say nothing till he speaks. As I took the 'phone from him, my nerves were at breaking point, but to my immense relief, he was very supportive and congratulated me on my success and hoped to see me soon, I just said thank you, like a little kid speaking to father Christmas. He didn't speak for long, thank God. He said he was ringing round all the successful candidates to speak to them in person. With a huge sigh, I put the 'phone down and poured out two classes of cognac. All we could do is sit tight and see what happens over the weekend. If we can get through till Monday we could be home and dry. But there's always someone to spoil things. It came with a 'phone call on Saturday afternoon whilst my husband was out shopping. I'd been expecting this call long before now, it was the political editor from the local press, the same guy that rang me the other week, wanting to know more than I wanted to tell him. What a time to ring, I was on my way upstairs for a lie down because I felt so wretched in every bone in my body. He said he'd seen the piece in last nights evening paper, and would like something for his own paper, for next week. I was on my own as my husband had gone out of town shopping, and I do mean "out of town" for a reason, it doesn't do to chance seeing someone at this late stage and getting caught up in conversation and letting something slip. The man on the 'phone was an experienced political editor, and not one to be patronised. He congratulated me on retaining my seat, ironically, he patronised me. Then went on to ask about my state of health. I don't know why I said what I said, but it's to late now. I must have thought we were invincible. Remembering my husband saying my voice sounded good on the 'phone, so I thought there's nothing for the reporter to get suspicious about as would be the case if I sounded weak and feeble as I actually am. I said to him, as he was only to well aware that I had a double mastectomy some nine years ago and made a full recovery, but I did have a bit of a scare last year, but after intense chemotherapy I was feeling a let better and hoped to be back in Westminster soon. Little did he know, it would only be to take the oath. He said it would be a sympathetic piece for me with a nice picture. After he rang off, I sat there for a few minutes and wondered if I had said the right thing, ah well, to late now. I hadn't been in bed long, well, not in bed, more, on the bed, when I heard the car pull into the drive, as I watched him through the window ferrying groceries from the car to the front door, I managed to get up and go down stairs to greet him, I still felt a bit groggy, but managed a smile for him, it was the least I could do. I made a cup of tea for us, as he put the shopping away and put his slippers on. As he sat down and asked how I was feeling, I couldn't help but shed a tear and said I feel I may have let him down, why on earth would you think that he said, as he gave me a gentle hug. After I explained what I said to our political editor friend he sat back in his chair and said we'll just have to keep our nerve. He didn't speak for a while and busied himself preparing something for tea, even though I wasn't hungry I did try to eat some of the omelette he made for me. After we settled for the evening, I managed to pluck up enough courage to try to say sorry if I dropped us in it, but did I need to say sorry, we both knew the risks we were taking, so I didn't' say anything for a few minutes, then he said sorry, sorry for what I said, well, er, not sure, but whatever it is, we sure as hell can't do anything about it now, we'll have to sit it out. We're to far up the creek with no spare paddle. So if it all goes pear shaped! tough! Sunday morning, dull, overcast, miserable. The weather wasn't much better either. As usual my husband was first up, it wasn't long before I could hear the creek of the stairs as he brought me the first of many a cups of tea. I drink a lot of tea, but I do like proper coffee now and again, it's the stronger taste I need, as my appetite, well, isn't really. I got up about ten o'clock and managed to have a shower on my own, it was my last little bit of independence and dignity I was determined to hang onto as long as I possibly could. Even though we have seen each other naked many times, I would not want him to see me looking so thin until he really has to, yes I know the longer I leave it the more difficult it will be for him, but I couldn't bring myself to let him see me just yet, but I'm sure he can imagine how I look. I went down stairs feeling a lot better, and sat by the window to try to feel the warmth of what bit of dappled sunshine there was. We always used to have a good selection of Sunday papers, as part of being an M.P. but not anymore, there didn't seem much point. We had a quiet morning doing nothing much, with only the odd 'phone call to congratulate me, my husband managed to make some sort of excuse for my absence to those who wanted to speak to me. Even though I was not too good physically at this time, I still had my wits about me, I said he should make a note of the excuses he made on my behalf, we didn't want to arouse suspicion by saying I was in the shower every time, it only needs one caller to speak to another, saying, when I rang, at one o'clock, she was in the shower, that's funny, because when I rang at half past three, she was in the shower then. We were a bit like tony Hancock puffing and sighing, Sunday for him was always a none day, not much on the tele", Sunday papers full of stories about the actress and the bishop, well, not quite, but you must have seen his show on the tele'. After some dinner, I felt a lot better and managed to tidy up, yes we woman always have to tidy up before we can sit down and relax, actually, we had a lot of planning to do, but where to start. How am I going to get to Westminster during the next week or so. As a reselected Member of Parliament it is imperative that I get down there to be made an "active" Member, but will I be strong enough to carry it off, if only for an hour or so. All we need is just enough time to take the oath, and sign the test roll, and get the hell out of there. By the way, I forgot about my treatment on Tuesday afternoon, not that it will do me any good as far as my cancer is concerned, as it's palliative radiotherapy, to make my condition more tolerable between now and, well, you know. The biggest hurdle is actually getting to London. Can you imagine it, putting my wheelchair in the back of the car, then onto the train, with the same fiasco at the other end. The journey alone, door to door would take the best part of three hours just to get there, then there's always the possibility of someone seeing me, or a photographer lurking around Westminster for a good photo' to sell. Once we were in London, what then, how do we get up all those stairs at our flat. I might have to hang around Parliament for a while before I was due to be sworn in, and of course we had to get back. We were looking at an epic journey that could take all and day and was fraught with danger. We did wonder if we could go down in the car the day before, but even that could be just as difficult, with the parking, and getting from car to flat. If we spent a night at our little flat there was always the danger I could take a turn for the worse. Not something I would want to leave my husband to deal with, the danger of not being in control of the situation was to much of a risk. But all is not lost, for we have a cunning plan. Whilst I'm having my radiotherapy, my husband is going to make some enquiries about hiring a private ambulance and nurse. I bet it'll cost a fortune, but it seems to be the only way. We went over and over all the pit-falls you know, the what ifs again and again. What if we had an accident on the way down, what if we got stuck in motorway roadworks, or lost in London, what if we ran out of fuel, what if my arse caught fire, there's just so much to contemplate, to many things could go wrong, but it's the only way. Well at least if we were running late we could always sound the klaxon. If it wasn't so serious it would be funny. Monday morning, the weather was much better, I managed to shower by myself, and felt pretty good, as I said earlier, some days are good and some days are bad, and so far this is looking like a good day. I was just negotiating the bottom stair when there was such a bang on the front door. I thought someone had thrown something at it, my husband was in the kitchen, making some breakfast, he came through to help me to the dinning room at the back of the house, we can't be too careful now. As he went to answer the door I was a nervous wreck, but far from throwing something, as he opened the door, the postman apologised for banging on the door and making such a noise with the sack of mail he had for us, he said there must be over two hundred letters and cards for us, wow! that was quick. Did I say wow. We hadn't opened any of them yet, I felt sure the many kind words of support and cards would continue, but nothing like this. There could of course be lots of negative comments about my absence from the hustings, or what about my state of health, and how soon will I be back at Westminster. We had just finished breakfast and was looking forward to opening some of the letters, when the 'phone rang again, my husband answered it, but was strangely quiet for a moment, I began to think all sorts of things. He said o.k. about twelve o'clock, we'll be ready, ready for what I thought. As he sat down I still wasn't sure what to think, neither was he, he said that was the area office, hell's bells, what do they want. My husband reassured me soon enough with a smile, and said the party leader is coming to our neck of the woods and did I want a photograph with him for the local paper. What did you say, well I could hardly say no. She said to expect an e-mail about ten to twelve, with the details of where to meet him, they had already contacted the local press, and they said they are willing to send a photographer as long as it was quick, and would send the same e-mail to each of us to say the exact time and place to meet them. What about my radiotherapy I don't want to miss it. He said that the timing, though tight, would just allow us time to have our picture taken, exchange a few words with the party leader, and skidaddle out of there, leaving enough time for your two o'clock appointment. So with our usual cup of tea, we had to work out a plan, a plan that on the face of it was a bit mean I suppose, but shit happens. We didn't want our agent to see me at a time like this, because I didn't exactly look a picture of health, so we hit upon a plan to 'phone him at the last minute, we knew he was busy at work, and couldn't just get the time off, as and when he felt like it, so it could just work out for us. With my usual trick with the wool frock, the lippy and blusher, and this time a woolly hat to cover my more apparent hair loss, and a leaf out of J.F.K.s book, we were ready to go. Quarter to twelve, I was just wondering if we had time for a quick brew, when the e-mail came through. We had just ten minutes to get there. My husband rang my agent, as luck had it, when he got through to his office, he was out seeing a client, perfect. I know it's a shame to deny my agent a meeting with the party boss after all the work he did for me, but we didn't want to put my agent in an awkward position. Besides we didn't want any problems with our tight schedule. We arrived at the prearranged location to see the photographer waiting in his car because of the rain, he just waved to us. I recognised him, but he didn't let on more than he had to, being local he knew the area well enough, and picked a good spot, not to busy with parking right in front of the shops. As we waited for the party leader, I watched the rain drizzle down the windows, a scene that reminded me all to graphically of the times we waited in that pot-holed car park all that time ago. After a few minutes a people carrier pulled up just behind us, it was the area campaign director for the party. As she waved, I could see the leader getting out of the car, as he shook hands with me and my husband, I could see the look of surprise on his face as we exchanged a few polite words. I hope he doesn't say anything to me about my state of health. We stood for a while, with the shops in the background as the photographer took a few pictures, and soon we were all done and dusted. We thanked the photographer, then as the party leader walked us to our car, he wished me well and said he hoped I would be back in Westminster very soon. As quickly as he arrived, he was gone. The photographer said one of the pictures will be in the next edition of the local paper.

And in no time at all, there we were, on our own with such a feeling of relief. I told my husband to get a move on as there was no time to loose, there is no way I'm going to miss my treatment. Why, two reasons, one needed it, two, and just as important, was the opportunity for us to make some enquiries about hiring an ambulance and a nurse. I told him to look on the Internet, but he said he would like speak to someone face to face. Once again it's off to that dreaded little green waiting room. Why do we always have the feeling we mustn't be late for a hospital appointment, then sit in the waiting room for such a long time. I'm glad we were the only ones in there and didn't have to wait to long. Even ten minutes in that sad little room, is ten minutes to long if you haven't got much time left. My husband left me as the nurse came to collect me for my treatment. He went to see if there was anyone to speak to about hiring an ambulance and a nurse. He said this is probably not the best place to start our enquiries, but we had to start somewhere. After all we hadn't made our minds up what to do about getting down to Westminster. If we had no luck here, he said, then he would look on the Internet. As I lay there on that firm hospital bed with the drugs gently pumping into me, I felt worse now than at any time during any of my previous treatments, even after my mastectomy all those years ago, I'm sure I didn't feel this bad. As I lay there I fell into a dreamy sleep, I dreamt a sort of joy, despair, light, dark, a real roller coaster of emotions. My escape from that dream was brought about by the nurse gently shaking my arm, as I woke and gathered my composure I could see my husband over the nurses shoulder, he smiled and gave me the thumbs up sign, as if to say everything went well with his information gathering. As he and the nurse helped me with my coat and shoes, we thanked the nurse, as the porter pushed me out to the car, we thanked him, and we were on our way home. Even though I was feeling very woozy and sickly, I couldn't wait to hear what he found out at the hospital, but he wouldn't budge till I was strapped in properly and we were well on our way home. After a few minutes I couldn't wait any longer and to my shame I got a little annoyed with him, but he is used to me by now and said wait till we get home, then I'll tell you all about it. With that I made a gesture as if to zip up my mouth, and closed my eyes till we got home and had a chance to gather our thoughts over a cup of tea. I couldn't wait a moment longer, and said come on then, what did they say about hiring an ambulance and a nurse. Well, he said, come on then I said impatiently, can we or can't we hire an ambulance. When we look back, we were a bit naive to think it would be quite that simple. It's not like hiring a car, but as luck had it, one of the nurse's nephew works for a company that does patient transport for the private sector. She gave him a card with all the details. She said we would need a letter from my oncologist or G.P. just to make sure I would be well enough to travel all that way, and how to get the appropriate nursing cover for the journey, Your doctor will probably recommend the Macmillan trust to provide a specialist nurse to accompany you as they are well qualified for this type of thing. We can arrange things over the 'phone with a minimum of twenty four hours notice. The bad news is, it'll cost an arm and a leg, about a thousand pounds to Westminster and back, plus something like two hundred pounds for a nurse, seems a lot I thought, but what the hell. We'll have to think it through right down to the last detail. On our way home we remembered the pile of letters waiting for us, and looked forward to opening some of them after my lie down, and wondered if any of them would be hostile. After my usual cup of tea, I had an hour on the bed with the curtains drawn and felt a lot better, and was ready to open some of the hundreds of letters. I couldn't help but be a little apprehensive, but we needn't have worried, we spent a pleasant hour or so reading dozens of very kind and encouraging letters and cards from supporters and well wishers alike, not one had any doubt in me or my ability to serve as their member or Parliament, not a negative comment in sight. Those reverential gags again, great!. After tea we tidied up all the letters and cards we had read, and put them in a couple carrier bags, and the rest we left in the royal mail bag for the time being, then it dawned on us, we did such a good job of convincing people, saying I'm "raring to go" that everyone believes I'm going to make a miraculous recovery like that of Lazarus, and we both know that's not going to happen. This is where it gets more and more nerve wracking, as we get nearer the time to chance the journey to Westminster and swear on oath to serve the Queen her heirs and successors etc, all it needs is one unguarded word either by me or my husband to ruin everything and not only cost us our dignity, and integrity, but as I see it, more importantly, two hundred thousand big ones. I know I said the whole package is nigh on a quarter of a million pounds, but I would get about thirty thousand pounds in any event. We will just have to measure our words and deeds very carefully from now on. My poor husband, he never once complained about his roll, or expressed his fears or misgivings, even though he's been in my shadow ever since I became an M.P. I suppose it's to late now to change things, and I feel sure he wouldn't have had it any other way, simply because he loves me. It's now Tuesday afternoon, just five days since my election, and once again we were both on edge when the 'phone rings, nor did we look forward to the postman coming, but if it's like yesterday we should have nothing to worry about. Once again I felt bloody awful, in fact I feel worse now than I have for a while, but we had to carry on.

THE PLAN

The next part of our plan, was to get the timing right to make the journey down to Westminster. What we need is someone to meet us there, to smooth the way and, well, not to put to finer point on it, fast track me into the commons, through the ceremony and out again so we can get back home with as little trauma as possible. We think the best day to go would be most likely be next Monday, there's only one way to find out. My husband would have to go down to Westminster to make the arrangements. Not a task he relished for one moment, those "what ifs" again. Rather than worry ourselves silly, we tried to put it to the back of our minds for an hour or two. It was only five o'clock, so I went to bed for a while. What a nice surprise when he came up stairs to wake me, I could smell my favourite, chicken curry and fried rice. He said he thought he would take a chance and nip out for some as a treat. Of course he said if I didn't want mine he could easily polish it off for me, but needless to say I scoffed the lot, it was superb. We did manage to forget our troubles for a while, as we watched our all time favourite musical, "Carousel" I did wonder once more if I might see 'Billy Bigalo' up there and polish the stars with him as I looked down on my boys. Then I thought, perhaps I was being a bit thoughtless watching this film, it's a favourite of my husband's too. He might feel he couldn't watch it without being reminded of nights like this. Then again, just maybe he would look up at the night sky, when all the stars in the sky were particularly shining bright and see me polishing away for all eternity, I hope so. Wednesday, another nice day, I waved him good-by as he went to catch the nine twenty three train. As he reversed down the drive he waved to me several times, he's going in his car today because he'll be back this evening. We only used a taxi if we were away all week. He soon disappeared out of sight, as I said a little prayer for him, yes I know, but I just want to clutch at any straw going. The steady stream of letters slowed to just a half a dozen or so each day. I must say it's been good to think just how many people supported me, but at the same time I'm going to let down just as many. All the people that helped me over the years with canvasing, leafleting, through thick and thin, summer and winter, but it's to late to turn the clock back now. My husband said he would catch a train back home as early as possible, and will give me a ring as soon as he knew for sure what time he would be home, even if it's late, he's not coming back without a strategy worked out. Needless to say he couldn't get back soon enough. I was determined to make an evening meal, even though my appetite not what it was, I just wanted to be his wife for as long as I could and sit with him at the table and have a meal together. It's getting hard work writing this down, I feel weaker by the day and sometimes catch myself wondering what might have been on that fateful day in October nineteen ninety five, if my lump had been benign. I suppose we can all think back and say what if this, or that. I pulled myself together, made some coffee and lay on the settee for a while, I thought to myself, I'll just close my eyes for a moment, by now, although I didn't tell my husband, I used to wonder sometimes, if I would ever wake up again. I slept for over three hours, in spite of the coffee. You know when you wake up after a daytime nap and wonder what time it is, or even what day it is. Well it was one of those days, but not to worry it was only just turned one o'clock. Now then whats for tea, or supper by the time he gets in. I raided the freezer and came up with some pork fillets, after thirty minutes on defrost, a few cloves, and a sprinkling of black pepper, and a couple of stock cubes stirred in some boiling water, and a slow cook in the oven, sounds good to me. He rang me about a quarter past four and said he was just about catch the train home, and expected to see me about seven o'clock. It was so frustrating that I couldn't even peel a few, potatoes, or even open a tin of peas or something, as my hands were to weak by now. The pork smelt scrumptious with the aroma of the cloves floating on the air. I watched and waited for the sound of his car, I'm glad I didn't have to wait too long, it was quite a strain looking through the window from my wheelchair. It was just turned seven as he drove up the drive, I do hope he has some good news. As he peeled some potatoes he said would I mind if he made them into chips as it was quicker to cook them, yes, anything, just tell me what happened at Westminster. Well, it couldn't have gone better, first, I went to your office, risky I know, but I needed to enlist the help of your office buddy, I had to take a chance that she was there, and she was. After the usual pleasantries and how you are, a question I found difficult to answer truthfully.

I explained briefly why I needed to speak to her, I said I would get back to her before I leave. A good move I thought to myself, she was the first real Parliamentary friend I had, plus the fact that she knows I'm ill, and has been very supportive over these last few months, and sending e-mails asking for help regarding her new health portfolio and keeping me informed about what was going on. She thought she was doing me a favour, but regular updates made me even more frustrated. I couldn't say anything to her, it would seem so ungrateful. She was one of the first people to congratulate me on the morning after my victory the other week. Now then, where were we, ho! yes, the arrangements for my trip to Westminster. He said the chief whip had got the e-mail we sent yesterday and had already had a word with the speaker's office, he said as I wouldn't be arriving till early afternoon, the best time to take the oath would be immediately after afternoon prayers, before the start of business, so far so good. Then he directed me to the office of the Serjeant at arms, to let them know what we wanted to do. As I explained the situation, they said they couldn't see any problems with my proposals, and went on to tell me the best way in. We should take the ambulance into New Palace Yard, and pull up along side the Star Chamber, just by the bronze plaque telling of it's history, it was known as the Star Chamber court, and it really was a court of law, built in 1602, There was a court there before then, Christopher Marlowe was tried there in 1593, but that's a bit before my time. We were to go into the passageway leading to the inner court, then up three steps on the left hand side, a few yards to the lift and up to the library corridor and eventually round to the back of the speaker's chair. Then, when the time comes, you can access the chamber between the opposition front benches and the commons table. We'll let security know when to expect you once you can confirm your time of arrival. If you can wait a few minutes we'll give you a letter of authority, just in case you need to explain things when you arrive at New Palace Yard. All I had to do was call the number they gave me half an hour before we are due to arrive, and would send a deputy down to meet us, couldn't be simpler. I then went back to see our friend, she said she would be more than happy to meet us on the day of our arrival. He told her he would send a txt message about half an hour before we were due, then again as we were nearing New Palace Yard. It may seem a bit odd I don't know all the Ins' and outs' of taking the oath, but it was a while ago and I was like a big school girl on my first day. So, if we can arrange for the ambulance to pick us up early next Monday morning, every one will have been sworn in by then, and hopefully little ole' me can slip in under the radar and out again, P.D.Q. So that seems to be sorted. Thursday, Four days to go. I hope I'm feeling better than I am right now. We had to work out a time schedule for the journey down, but where to start, it was like one of those kid's conundrums, if the ambulance's average speed is fifty miles an hour, how long will it take to travel two hundred miles. On the face of it, one would say four hours. It's not as simple as that. My husband takes the best part of four hours in his luxury fast car. After several cup of tea we were no nearer the answer, so we were back where we started. It's amazing how the blindingly obvious can sometimes escape you. The ambulance hire people will know how long it will take.

So we rang them right away to put us out of our misery. They told us, whilst they could not guarantee how long it would take, as there are a number of imponderables, but thought five hours would be a reasonable time, but to allow time for the possibility of any hold ups, so we booked an ambulance right away, for seven thirty the following Monday morning, and asked my doctor to arrange for a Macmillan nurse to accompany me on the journey. We knew we had to leave early on Monday morning, to be able to make it to Westminster before the afternoon session begins about quarter past two. We don't want to be to soon, other wise we'll have time on our hands. We had our little bolt hole of course. We could use it in an emergency, but that was the last thing we wanted, and what about the ambulance, the nurse, and the driver. So we agreed to go early and just hope we didn't run into any problems on our way down. Last week, the local news reporter, said he would do a sympathetic piece in the paper, with one of the pictures we gave him. I'd forgotten all about it until my husband gave me the paper from his brief case, he'd forgotten as well. It was a smashing piece, the reporter really did me proud. So, Monday it is! Four days. If I was fit we could have a few days break, then again, if I was fit, I wouldn't need a break would I. We did have a ride out on Thursday afternoon, we pulled up in the car park of a stately home miles from anywhere and ate ice cream as we reminisced about our life together. I couldn't help but recall the words of a favourite song of mine. "The wind beneath my wings". This is my favourite verse. "So I was the one with all the glory, While you were the one with all the strain. A beautiful face without a name for so long. A beautiful smile to hide the pain. Did you ever know that you're my hero"...... I'll stop there or I'll begin to cry. Pleasant as it was, I was glad to get home, and yes, you've quest it, have a nice cup of tea. The sickness had subsided by now, and I felt a lot better as my aches and pains became a little more tolerable as we sat by the fire and watched a bit of tele', my appetite had improved a little bit by now, it must be the fresh air. I hope I can keep this level of well being till next Monday at least. As I said, my appetite had improved, so I asked his lordship, if he fancied a fish supper, good idea he said. I asked him to get me a fish without batter and only a small portion of chips, o.k. he said and was gone in a flash. We sat and watched the evening news with our fish and chips on a tray in our laps with a nice fresh muffin each, smashing. Thursday night on the tele' isn't to bad, with the police and hospital drama's. You may think it strange but I still like to watch hospital drama's even though they might feature something similar to my own illness, but, well, there you are. We used to watch question time when ever could in our little flat in London, it seems years ago now, but, it's only about eight months ago. But not anymore, there doesn't seem much point, my heart wasn't in it. This "there doesn't seem any point", is to become a recurring theme from now on. To be honest, I've got used to the idea of shuffling off this mortal coil. So long as I can secure the future for my husband and the boys. About ten thirty we were both feeling tired, so off to bed we went. We still sleep together, I couldn't ask my husband to suffer the final indignity of sleeping in one of the boys rooms, I suppose it will happen soon enough if I end up in hospital for the final chapter. Friday, more like cry day, remember a week last Monday when I said that I had never felt such pain, well , I am beginning to feel just as bad as on that awful day. My arms, legs, back, and now my chest feels like a heavy weight is pressing down on me. I hope it eases during the next day or two. There's only just over forty eight hours to go, and I feel worse than ever. I'll tell you while he's out having his hair cut, if I feel like this on Monday, we'll have to call it off. What a sickener that would be. Whilst he's at the hairdressers, he's going to ask one of the girls to call round this evening to try to work a miracle on my hair, I used to have such a good head of hair, but it's no use dwelling on what was. Just let me tell you this, at first when I was ill, I woke up to find clumps of hair on my pillow, and believe me it really brings it home just how powerful those drugs are. Yes, I cried then too, but this time the palliative therapy doesn't cause the hair loss as before.

GOOD PRACTICE

I was really apprehensive when my husband came home with his smart haircut, and news that a young lady was calling round about seven o'clock to trim my hair. He said he thought the girls might be eager to get off home, to get ready for their night out clubbing. Have they nothing better to spend their money on I thought, he went on to tell me she was a young mum and needs the money. As the hands of the clock moved slowly towards seven I began to feel very uneasy and couldn't even think of having anything to eat till after the hairdresser has gone home. About ten to seven, I went up stairs to do my usual trick with the wool frock, a bit of lippy and blusher. There was nothing more I could do, I'll just have to front it out and put on a brave face during the time she spends with me. It could be good practise for Monday, and no doubt my husband thought so too. We replaced the mirror over the fireplace, with a large picture called "The Gleaners" by John-Francois Millet, just swanking, we didn't want her to see me from to many angles. Then we put a chair directly under the centre light so there would be enough light to work by. We prepared things as much as we could, but what the hell, we'll just have to wing it, as they say. A pretty girl, the hairdresser, about twenty one or two, and fortunately hadn't a clue who I was, and probably didn't even know that there's just been a general election. There's a lot to be said for the lack of interest in politics by the young, well, on this occasion at least, as she washed my hair over the kitchen sink. How I managed to walk across the room I'll never know. As I leaned over the bowl my back was killing me, she must have realised that there was something wrong with me, but she just carried on without comment, and made me look presentable with a well placed curling brush. She hardly said more than half a dozen words, she even refused a cup of tea, there was a time that would never do in this house. I guess the reverential gag came out again. I just hope that there's an abundance of reverential gags next Monday. After she'd gone, it was getting on for eight o'clock, and I did feel like a bit of something to eat by now, no, not a curry, chicken soup was about all I could manage, some people call it "Jewish penicillin" it was very good. I can't remember what my husband had, I was to tired to notice. When I finished my soup I dozed off for a while, I try not to sleep too much during the day so I can sleep better a night. I tell my husband not to let me sleep to long during the day, but he finds it difficult to wake me at times, so he lets me sleep on. I'm sure he thinks if I'm asleep, I'm pain free. I suppose I am really. Saturday was quiet, he must have been very careful not to wake me as he got out of bed, and let me have a good lie in. I'm glad he did, because I forgot to tell you, this will be the first time I've seen the boys since before the election. As I said before, they helped a bit during the election just to show willing, for their dear old mum. Now they have their own lives to lead with their teaching jobs and new friends, no girls yet! I do hope they like girls. They are due about noon for a flying visit. They're coming up to attend the wedding of a former University friend, and expect to see a lot more pals from uni'. So with my new "hair do" and the usual wool frock, I though I looked quite presentable, as dad opened the door and let them in I couldn't help but notice they were a little shocked to see me standing by my wheelchair, but then again what can we tell them over the 'phone, all I could do was try to reassure them that I, or rather their dad and me had come to terms with the prospect of me "going" sooner, rather than later, not something a mother wants to tell her children at any age, but that's the way the cookie crumbles. I told them we have hired a private ambulance to take us to Westminster on Monday. I said we were going to go in the car but thought better of it. I said I had to sign in as it were, I didn't mention taking the oath or anything they didn't need to know any more than that, I know they must remember my taking the oath some four years ago, but they knew nothing about the financial implications, or the difficulties we could face if we were to be asked any awkward questions. Probably for the best, they might not approve, in fact I'm damn sure they wouldn't. It was such nice sunny day as we sat in the garden for a while as they told us about their new teaching posts, and how they were getting on, I was so proud of them. It felt good to be waited on by three men, they made a lovely lunch with salmon and fresh salad with some nice firm tomatoes, and soft bread just for me. We talked a while, but it was soon time for them to get off to their freinds wedding at a nearby church for three o'clock. A flying visit I know, after hugs all round and fighting to hold back to tears, we waved good by to the best sons any mother could wish for and thought to myself, would I ever see them again, then had the horrible realisation that I probably wouldn't. A thought, that really struck home for the first time since I became ill all those years ago. But you never know, me and Billy Bigalo might be watching over them as we sit up there polishing the stars. I'll pick three big bright ones for my boys and polish them for all eternity. I went to bed that night with a smile, not a tear. Twenty four hours to go, blimey, I'm beginning to panic now, you know those "what ifs" if I'm this bad now, how about tomorrow. The ambulance was due at about seven thirty, we were told to be ready in plenty of time as private ambulances weren't allowed to exceed the speed limit, unless it was a real emergency. That was something we figured out already. To my surprise, when my husband jokingly said to the patient transport people, we can always use the siren if we were running late, they were not amused, as it is strictly verboten. My stomach felt as if it was full of bricks, a feeling I've not had since I was carrying the twins. I spent a very restless afternoon on the bed feeling just awful, my husband brought me a cup of tea about four o'clock for which I was very grateful. He said I didn't look too well as he sat on the edge of the bed and held the saucer under my chin. I said I feel as weak now, than at any time during my illness, it must have been all the excitement yesterday. I know he didn't want to put me through more anguish if at all possible, but felt he had to ask if I thought I make it tomorrow. The truth is, I had barely the strength to hold my saucer let alone think about tomorrow, we'll just have to wait and see.

ARRRGH!! WHAT A BUMMER

We didn't have to wait long, sometime around three o'clock in the morning I really thought I was dying, and to my shame I wished I had. My husband 'phoned the emergency doctor, but by the time he arrived I'd been sick a couple of times and was beginning to feel a little better even though I still had the shivers. With words of reassurance and comfort only a doctor knows how, and a shot of morphine I soon felt better. We had to tell him we hoped to go to London tomorrow, but alas he said in the strongest possible terms that it would be a suicide mission, or at the very least, shorten my life by two or three weeks. He advised against making such a journey, even by private ambulance, at least till I felt a bit stronger, maybe in a few days, or even next week. Then he asked the most difficult question possible. Have you guessed what it was, yes that's right. Why do you need to go down to Westminster at this particular time whilst you are so ill. Well ! all we could say is that it is a legal requirement to sign in person as soon as possible. I'm not sure he understood, but it seemed to satisfy him alright, so he bade us good night, or rather good morning and said he or someone would call tomorrow. I must say, even though it was a bitter blow, we had to agree with him, it would be a suicide mission. I don't remember the doctor leaving the house I was oblivious by then. The next morning was a beautiful day, as my husband came up with my cup of tea, he said he'd been up since six o'clock, I tried to sit up and soon realised that there was no way I could have made such a trip today, I felt as weak as a kitten. What I didn't know, was that my husband had arranged for our two sons to meet us at Westminster but had to 'phone them, to tell them I will not be going down today, and not to worry, as we felt sure we would see them next week. He had also 'phoned the ambulance people to cancel our trip, at some cost I might add. Then sent an e-mail to my friend who was going to meet us when we arrived, he also sent an e-mail to the chief whips office to explain the situation and let the Serjeant-at-arms know what the situation is, and we would be in touch, and would make arrangements for next Monday. After a second cuppa' I went down stairs, not showered, or even dressed. I just sat there with the big blue dressing gown wrapped round me and starred at the carpet, counting the little swirls on it, for the first time in weeks, I actually felt sorry for myself. When I looked up at the clock I was surprised to see it was only ten o'clock. All I could see, was a long, long, day stretching out before me. I was not a happy bunny, so I had another cup of tea then dozed off again. My poor husband must have been creeping about all morning so I could get some rest. You may wonder why I didn't stay in bed, well, would you. About one o'clock he offered to make me something to eat, I nearly said no thanks, then said I would like a cheese sandwich with a few dribbles of

Worcestershire sauce, and a strong coffee, it will help pass the time, and give him a feeling of being useful. He grated the cheese, and cut the crusts off the bread, how's that for a loving husband, I didn't enjoy the sandwich very much, but managed to eat it all. The coffee was very good, with a nice rich taste. I drifted off again and from time to time I could here the television, but had no idea what was on, it's funny how when something is turned off, the silence can actually wake you, that's what happened to me about ten 'o'clock as my husband suggested we had an early night, no, not that sort of early night. Speaking of which, we haven't had an "early night" for months, and I don't think we will ever again, better shut up now or I'll set myself off crying again. Believe it or not I had a good nights sleep, and feel fine this morning, even though I slept most of yesterday. The only thing to spoil my day was the frustration to think that I could be an "active member" by now, and the pressure would be off, not just off, gone! Home and dry! But like I said before, not much we can do about it now. I mooched about in my wheelchair, sighing, looking out of the window. Tuesday was just another day to endure, not just for me, but my poor long suffering husband. We live at the end of a cul de sac, a setting that suited us when we first moved to this house with nice long gardens for the boys to have a football kick-about with their dad. It is even long enough to have a bit of bowling prictice in the summer. But on days like this I wish we were situated on a through road, with a bit more activity, it's so quiet here, ah well. You can't credit it can you, next morning, Wednesday I felt even better, my husband wondered if he should go to Westminster while I was o.k. to be on my own, and we still had the chance to put things back on track. He had plenty of time to catch the 11. 23 train, he would be there by two'ish. He could nip to my office, grab my post bag, see my friend, or leave a note if she isn't there and be out again in under a hour. Whilst he's down there he could go to our flat, sort out the through the mail, and make sure that things were o.k. Even though I'm a new, or re'elected member of Parliament, I cannot act officially for my constituents till I have taken the oath of office. I still get mail sent to me at the commons as usual. The senders don't know about my status, or lack of it, not being an "active" M.P. As mail is addressed to me, no one else can open it anyway. So off he went with not such a cheery demeanour as before, but he was hopeful he could put things back on track for next weeks trip, or at least have a damn good try. There isn't a time limit to take the oath, and sign the test roll, but it does mean as a non "active" member, I cannot use my office, and don't get paid after the current month of the election, but for me, there is a personal time limit! The limit dictated by my state of health. I know I've told you before, but I have got to sign on the dotted line before I can collect my gratuity, sounds simple, but can we pull it off. After a quiet morning of doing nothing much. I got his expected call, about two o'clock, we let it ring out three times then wait a minute, then he rings again, we don't want any unwelcome 'phone calls do we. He said he would be catching the four o'clock train and would be home about six thirty, can't wait I thought, how did it go I asked impatiently, every thing's fine you'll have to wait till I get home he said. I was looking forward to him coming home more than usual because I managed to make a steak and kidney pie for tea. well, made a start, even though I wasn't very hungry at the time. I felt I had to try to make an effort. How did I get it in the oven you ask, well, I defrosted some steak and kidney o.k. and chopped a few onions into chunks as best I could on a big chopping board on the kitchen table, then I put an empty dish in the oven, and spooned some of the meat and onions into a sauce pan so I could carry it, and very gingerly managed to get across to the oven and pour in the first bit of meat and onions in the dish, then did the same a couple of more times, turned the gas down to mark four, got back to the kitchen table, and sat down totally knackered. Then I realised I hadn't put any stock cubes or salt and pepper in it. My hands were to weak to crumble the oxos so I bashed them with a rolling pin to break them down so I could stir in some boiling water with the salt and pepper. Ten minutes later I finished the casserole. I then remembered to take some jus-roll pastry out of the freezer to defrost, and sat down again truly, truly, beat. When I said not quite made a steak and kidney pie, because, when he came home I would need to ask him to just roll out the pastry, jus-roll, get it, oh, please yourselves. It was well after half past six as I saw him coming up the drive in what seemed a cheery mood, as I could see his broad smile through the windscreen, and guessed his trip was a success. The journey down was uneventful, he went through security without any problems. Then he went to my office, collected my post, and left a message for my colleague to say he would 'phone her during the week. Now for the tricky bit, he went to the "Whips" office with a certain amount of trepidation, he thought it would be common courtesy to tell him of my intended trip down next Monday to take the oath, and yet at the same time, begged and preyed that he wouldn't be there to ask difficult questions, and as luck had it, there was just a secretary on duty, it's a good job he had the forethought to write a fairly formal letter to apologise for my absence, and to say I would be taking the oath on Monday next. He handed it to the girl and promptly left, hailed a taxi, caught the train, sat down and breathed a huge sigh of relief. As he rolled out the pastry and covered the pie, he peeled a few potatoes, then opened a tin of peas, "eat your heart out Gordon" and very nice it was too. It was nearly eight by the time we had eaten our tea. Then he filled me in with the finer details of his trip. He didn't see anyone of note, as there was a debate going on. Oh how I wish I was there. After a cup of tea we went through the post together. Most of it was routine Parliamentary stuff. The rest was from our flat, and, for the most part was junk mail, there's no escaping it is there. And so to bed. I woke up to a cup of tea as usual, on such a lovely day and feeling of optimism, and I must say I felt much better, not particularly great, but the nice weather helps. Now that I am feeling better, I was determined to keep myself in shape for Monday. It could well be my last chance. During consultations with my oncologist over many months, reading books, plus my knowledge of human anatomy, I knew that the latter stages of cancer sees a rapid decline in my body mass, once I had started to develop the condition know as cachexia, or muscle wasting to you, the weight loss is so debilitating it would make me feel very fragile indeed. I have lost well over two and a half stones over the past three months and feel as week as a kitten at times, and unable to stand for more than a few minutes, let alone walk. So, it's rest, medication, and more rest till Monday. Friday was a day I used to look forward to when we were in London together, because sometimes we would ask someone to take my surgery for me so we could stay in London for the weekend and go to the theatre, or a posh meal somewhere, but I mustn't dwell. Today is another beautiful day, after my usual cup of tea, or should I say cups of tea, I tried to have shower on my own, but alas I had to ask my husband to help me. Much as It pains me to let him see me like this, so thin, not the voluptuous woman he married. I couldn't take a chance on my own, for fear I might have a fall and put paid to everything. We sat on the edge of the bed whilst he towelled me dry, and helped me get dressed. As I said, it was a lovely day, with the sun shinning down on us, as we spent a very nice quiet morning watching a colony of ants scurrying from one crack to another round the patio, totally oblivious of us watching them. He said I'll patch up those cracks up when you've g .... he turned and fell to him knees, with his voice quivering, asking my forgiveness, he really didn't need to, I could never accuse him of being thoughtless, there was no need for any words, we just held each other tight. Even though I didn't feel tired at all, I thought it would be a good idea to have a rest for a while so I could watch a bit of tele' this evening without falling asleep. So with a dose of my medicine and the curtains drawn I had a good three hours of not so much sleep, but more, a good rest. I haven't said much about my medication up to now, for if I did, I wouldn't have time to tell you anything else. Needless to say I was on regular visits to the hospital for my palliative radiotherapy and a supply of morphine tablets to help me cope with the pain on a day to day basis, and pills to stop me feeling sick. But all in all I don't feel too bad for someone on their last legs, even though I would like to live a little longer, I just hope I can make it o.k. on Monday. We'll just have to hope for the best, and a nice day of course. The weather is being very kind to us again today, Saturday. How we keep getting away with no one calling, or 'phoning me, I don't know. All's quiet on the western front as they say. I made no excuses regarding my constituency surgery, my agent made no attempt to contact me, either by 'phone or e-mail, and not a word from head office. Suffice to say, my husband did the shopping, but waited 'till near dinner time before he went out, just in case anyone called. I did nothing much, just waited for Monday. Though I'm feeling o.k. at the moment, I still feel apprehensive about the journey down to London, it's a long way back if anything goes wrong. The trick is to look a bit heavier than the seven and a half stone I actually am. It may seem a bit of an extravagance to you, but it is an investment, have you guessed yet, you have, yes a posh new frock, where from you may ask, would you believe, Harrods, what a lovely gesture. Remember when I said I might actually buy something from Harrods before I die. He should have got it from Marks and Spencer's, then he could have taken it back later. Pure wool, dark green with buttons down the front, and a high collar. What a lovely man. Sunday, twenty four hours to go before the most important day in my life. I just want to be able to stay focused for half an hour or so, that's all I'll need to read the oath in a good strong voice, bow to the speaker, sign test roll, and leave. Sounds easy doesn't it. With the ambulance and Macmillan nurse booked for the second time, we must make it tomorrow. With my medication in one bag, my makeup in another, and a complete change of clothes including underwear in a third. There was nothing more to do but prey, yes I know what your thinking, but you never know, I might meet Him or Her soon enough, then I would look a fool. We sat the alarm for six o'clock, and hoped for good weather tomorrow.

COULD THIS BE A GOOD DAY?

As my husband brought me a cup of tea and pulled back the curtains so I could see the clear blue sky even at such an early hour, we never close the main curtains, we like let the morning sunshine in as early as possible. He said it looks like it's going to be a lovely day for us. And a long one I thought for a moment, you know the "what ifs", Well here goes, I wriggled my toes, o.k. then I flexed my legs good, made a fist o.k. lifted my knees up towards my chin, very good, then breathed in as deep as I could, a bit of a cough, but that was o.k. too, then I slowly sat up. To my overwhelming relief, I felt pretty good, but wait a minute, I haven't tried to stand up yet, I've been using my wheelchair quite a lot these last few weeks, the problem with that is, it's easy to get more dependent on it. Well, here goes, as I swung my legs over the side of the bed and slowly put my full seven and a half stone on them, and stand up, they held up very well. I managed to walk to the window and look out just to make sure it really is going to be a beautiful day for us, in more ways than one. As my husband came in to see if I was ready to get up, he automatically looks at the bed, but had the surprise of his life to see me quite upright and seemingly as fit as I have for quite some time. I was so pleased to be able to greet him in a standing position for the first time in about three days. We both said, at exactly the same time, come on, no time to waste! and actually had a laugh together, another first for a while. All I could eat were a few soggy cornflakes, my teeth are to tender now to bite into some nice crunchy toast and marmalade, but my husband did eat a hearty breakfast. All this before seven o'clock. We sat in the lounge for a while, no more tea for me, I didn't want to have to wee to often on the journey down, but my husband had another cup as he tried to comb my hair, I said I would ask the nurse to do it for me, hope it's a woman. About ten past seven, we heard a car door bang, and thought the ambulance was arriving a bit early, but to our horror, as we looked through the window we saw the worst possible sight we could have wished for, The press! Shit! how did they find out. Don't ask me he said with ashen face and not a little panic. I'd better go and offer them some tea or coffee and try to find out where they're from, and how did they knew about our trip to London this morning. A bacon butty wouldn't go amiss, I said, but we didn't have the time. Besides one of the could be a Muslim, or Jewish, or, or , or, pull yourself together woman! When he came back in he said there are only three of them, one cameraman, one sound man and an interviewer or someone, an interviewer! I can't speak to anyone, it could spell disaster for us. He asked them how they knew we were going down to Westminster today, of all days, but they either couldn't' or wouldn't say.

However, they did give us a clue. the cameraman was booked to come here last Monday, and said it was cancelled at the last minute, the only people that knew of the cancellation last week were the hire people and the driver. I don't think I'll say anything to him it might be a different driver. I don't think it will be so bad, they only want a couple of pictures and a quick word, we'll just have to tough it out. As we were wondering how to handle this very awkward situation, there was a knock on the door, my husband opened it to a very pretty nurse with lovely red hair and a beaming smile, even at seven thirty in the morning, makes you sick doesn't it. As I looked through the window I could see a gleaming white ambulance with light blue and dark blue chequer patterns down the sides. It wasn't exactly what I was expecting, though I'm not sure just what I expected. My husband said it was brand new, and it's capable of a damn sight more that fifty miles an hour. Thats great, he said, we'll be there in no time, in fact we might need to spend some time at a motorway service station so we won't be to early. We, or rather my husband, made the 'phone calls to the boys as promised, then rang my colleague in London, who said she would meet us in New Palace Yard. I know he didn't need to ring her till we were nearly there, but it's better to be sure. As I said, this could well be our last chance to get to Westminster. The ambulance driver and the nurse came up the drive and into the house with a manual wheelchair, you know the type with four equal sized wheels, even that was brand new, the nurse was carrying a big red blanket over her arm, as the driver helped me into the wheelchair with ease, the nurse was just about wrap the blanket round me, but I said I was more than warm enough thank you. What she didn't realise was that not only would it highlight my condition, but was I already sweating with apprehension as to what was waiting for us at the end of the drive. I'm glad the ambulance driver was a big chap, he turned me round as if I were a child in a push chair, and once though the door, he stated off down the drive like a man possessed, with the pretty little nurse scurrying behind. With my hair brushed, and my new wool frock under a not too heavy coat and floppy hat I didn't look bad at all, and felt better than of late. The news team said "just a couple of pictures please" a couple of pictures my arse! well, not exactly, but you know what I mean. As we approached the ambulance and the "gentlemen of the press" I put on a brave face, and a big smile, and said good morning gentlemen, what lovely day it's going to be. It was at this point that we realised that the camera was a video camera, not a stills camera as my husband thought it was. How stupid to even think it would be anything but a video camera, to cap it all my husband asked if it was a live broadcast, how could it be live, they didn't know what time we would be coming out of the house, or did they. It seems we panicked and almost blew it a big time at the very first hurdle. They said they were recording, and didn't know when, or even if, it would be used. We didn't believe a word of it of course, but what could we do, we could see they were from the local television news as they scurried around for pictures from all angles, we asked the nurse and driver to keep moving as we don't want any hold ups, but the reporter was a brute of a man, he elbowed the pretty little nurse in her breast and she nearly fell backwards onto the garden. I cringed as I realised just what damage that could cause years down the line. With not a word of apology to the poor girl, he asked me about my cancer, how did he know about my cancer I thought, till the penny dropped, the ambulance driver of course, he was only one of about half a dozen people that knew my journey was cancelled last week, and that we had hired a specialist cancer nurse. By now I was really beginning to feel uncomfortable, half in and half out of the ambulance with the reporter asking awkward questions, with a microphone in my face. I was really beginning to panic now, I tried to play it down, I was so damned annoyed with him pushing the nurse like that, but he insisted I gave him some sort of comment. We did our best not to be rude and tell him to bugger off, as I dearly would have liked to, but the reporter kept on at me, with the cameraman in front of me, the sound man far to close for comfort. Then it happened. the reporter asked me point blank, " how long have you know that your cancer had returned, and just how bad is it". No reverential gags here then. To my shame I lost my composure and blurted out. So I've got cancer, what do you want me to do about it, put up a banner, saying I've got cancer! With that I was pushed unceremoniously into the ambulance and the doors shut behind us. My husband got in the front and was able to come through to the back and sit with me till the news men had gone. My husband was going in his own car, and would stay close behind us just in case there were any problems. He really didn't want to take his own car down the motorway behind the ambulance, at fifty to sixty miles an hour, but what else could he do. The nurse told me not to worry, she was a specialist cancer nurse and was well qualified to administer any appropriate drugs as necessary, I thought thank God for Douglas Macmillan. She had some of my medical notes from my G.P. just to be sure we had all the bases covered. After she had made me comfortable and familiarised herself with my medical needs, she said to the driver she was happy to go. My husband gave me a kiss and then spoke to the driver about our first stop on the motorway, the M.1, I think. We didn't' know at the time but the driver, worked as an ambulance driver in London, based at St' James's, or Jimmy's as it's affectionately known, and knew the place like the back of his hand, good choice of driver bat man! He said to my husband, the problem is, we won't be allowed to stop outside the main entrance at Westminster, as it was not an emergency, he would have to take us round the back, at that point, my husband quickly put his mind at ease, and said he had made arrangements for us. All he had to do was turn right into St' Margaret's street, and left into New Palace Yard and someone will be there to meet us, good he said, and we were off. We stopped a couple of times on the way down, but I can't recall much of anything really. I couldn't help but doze off with the drone of the engine, and the hum of the tyres I think the nurse was finding it hard to stay awake on such a warm day. The driver popped his head round the partition to let us know we were about twenty minutes away from Westminster, the nurse shook my arm a little, she said we had better get ready for our arrival. I had no idea where we were, I couldn't see through the frosted glass windows of the ambulance, only to see it's a nice sunny day out there. My husband had txt my colleague, and rang the 'Serjeant at arms' office to let them know we would soon be there, and to remind them that they said they would send a deputy to meet us when we arrive. We turned into New Palace Yard, along the side of the Star Chamber and pulled to a gentle stop exactly by the plaque I told you about earlier. With my husband close behind, the nurse helped me to get properly dressed and brushed my wisps of hair into something resembling the way it looked last week, and I didn't feel to bad at all. My husband knocked on the door, I don't know why, he's seen it all before, as he stepped in to see how I was, I could see he was pleased with the way I looked. When the nurse and the driver stepped outside to stretch their legs and have a look round, he took the opportunity to whisper to me that the place could well be covered by security cameras, and said it's important that I present myself as well as I can. When the nurse came back in to see if I was ready, she said the ambulance driver knew one of the security guards from way back, they were both ambulance drivers based at St James'. As I stood up I felt a bit wobbly for a moment and began to fear the worst, but gradually, like one of those reptiles you see on the tele' I began to feel a bit better as I seemed to warm up in the sunshine as each moment past. We still had our passes from the previous administration, so we would be o.k. with them for the time being. As we thought, there were no problems at all. One of the 'Serjeant at Arms' deputies greeted us and showed us to the lift. Our friend and colleague came round the corner to greet us, she too had to show her pass because she was not known to these Particular security men. I suppose it's better to be safe than sorry. We told the driver we would be back in about half an hour. It was twenty past two, perfect timing. I managed to get myself into the wheelchair alright, because there was no way I was going to risk walking to the chamber, after all, I wouldn't be the first person to enter the house of commons chamber in a wheelchair. My colleague from our office, with whom my husband made the arrangements had asked a male colleague to help us.

He greeted us just by steps leading to the lift. I was very happy to see him as he was the first person I got to know on my first day hear, remember, my mentor for the first few weeks as a new member. He and one of the security guards lifted me up the few steps into the passage and along to the lift. The security guard and the deputy Serjeant at Arms wished me well as the lift doors closed. We began to ascend to the library corridor, I was beginning to get very nervous by now. We trundled our way along the corridor behind our very own guide or usher or whatever he's called, we chatted quietly about the journey down as we approached the back of the speaker's chair. The usher told us our sons were already on their way up to the visitor's gallery, my friend has arranged things for us very well I thought, as my husband gave me a kiss, and whispered in my ear, all you have to do is stay focused for ten minutes and were home and dry. He then went to join the boys in the gallery. Then, to my horror, my colleagues said they were going to take their seats on the benches, and someone else would come to help me. I must say, by now my pulse was about two hundred beats a minute, sweat was pouring off me, I thought how can I possibly manage on my own if this other person doesn't turn up. This is one 'what if' we hadn't bargained for. As I waited behind the Speaker's chair just outside the chamber and just about to wet myself, pulse racing more than ever, with nowhere to turn, and just wanting to get the hell out of there, then to my immense relief I heard someone come up behind me and whisper in my ear, you'll be o.k. I'll help if you get stuck, it seems he knew I was taking the oath today, but had only just been told I was waiting. As I looked round I was greeted by one of my closet friends, not only in the house, but socially as well. When we were in London he and his wife dined out with us on many occasions, and often went to the theatre together, but alas, that was in the good old days. It didn't bare thinking about what might have happened if he hadn't come to help me. As the time drew closer for me to cross the Rubicon, for that's exactly what it felt like, the point of no return. I began to realise just how much I've missed this place, and began to shed a tear as my thoughts go back to all the arguments, the points of order, the amendments, white papers, briefings, and best of all, the nights out in London, with the theatre, restaurants, the whole place seemed to be alive twenty four hours a day, but no more. I know I will never see this place again. As I sat there it went very quiet. I wasn't sure if this was a good thing or not. I could just hear Mr Speaker coming to the end of prayers before commencement of the afternoon session and wondered should I confess to my friend, and just as importantly, to myself. Should I tell him I might not be alive in a few weeks let alone make it to Parliament as a member. I particularly respected and grew to admire him over the years, and didn't want to get him in any trouble. As I glanced up to the gallery, I could see my two sons watching and waiting for me to take the oath and felt sure they were very proud of their dear mum, and totally oblivious of the consequences if anyone was to cry foul. My husband, sat next to them, probably just as much a nervous wreck as I am right now, thinking of the gut wrenching embarrassment that would surely follow if I was bared from taking the oath. I'm ashamed to say even at this late stage, the money was a very powerful incentive.

I was determined to go through with it come hell or high water. I could just hear Mr Speaker closing afternoon prayers as my heart began to race up to a zillion beats a minute. With nerves jangling like those of a crocodile in a handbag factory, I began to think why am I doing this, putting myself though this torture. I was on the very edge of panic. My adrenaline kicked in for a few seconds, and I thought lets get on with it. I came back down to earth just as my companion said well, hear goes, good luck. It was at that precise moment I realised he must have known I was very ill indeed. He could get himself in a lot of trouble if it became known that he was party to my deception. I told myself to tough it out for the boy's sake. All this happened in a flash, I even said a little prey as we crossed this imaginary Rubicon at the side of the Speaker's chair, and the point of no return. As my companion wheeled me round the side of the Speaker's chair and into the chamber proper between the commons table and the opposition front benches. There was just enough room to pass through as the shadow cabinet were not in their seats yet. The distance from the side of the Speaker's chair to the front end of the commons table is about five to six yards, but it might as well have been a mile, I was so nervous. Even though I didn't want to look to my left or right out of shear fright, I did acknowledge my friends and colleagues of my own party. My "chauffeur" as it were, told me that all my colleagues had been paged to come to the chamber to see me taking the oath, not least, our leader who was such an inspiration in the early days of my campaign, he knew more than most that I would be looking decidedly poorly by now. I could hear some of the members murmuring their surprise, or was it sympathy, or approval, or even disapproval. I wasn't sure. What if someone did cry foul. I looked at the floor 'till we were almost in front of the commons table. With sweat pouring off me, my heart about to burst, words can't describe how I felt as we came to rest inches away from the "mace", right there in front of me. As I stared at it I was mesmerised by it's splendour. I've never really taken in what a potent symbol of authority it is. I thought, this is it, the moment we've been waiting for. The clerk to the commons table leaned over to pick up the Bible and put it in my right hand. With nerves at braking point, I was glad I memorised the oath as my eyes were beginning to fail me. I took a deep breath and surprised myself with my voice as I recited the oath." I Jacqueline James, swear by almighty God that I will be faithful and bear true allegiance to Her Majesty Queen Elizabeth. Her heirs and successors, according to law. So help me God", Then to my horror the Speaker raised his hand for a second and stepped down from his chair, what the hell dose he want I said to myself as he came round the table to speak to me. I thought he was going to question my state of health before I signed the test roll, another minute he wouldn't need to ask, I'd be in a state of cardiac arrest. If I couldn't sign the test roll we would be in deep dodo. The signing was vital as you well know. Then, to my immense relief, as he leaned forward he smiled and took my hand and wished me a speedy recovery, he handed me the test roll or parchment book, as it is now, then he handed me his fountain pen to sign the book and enter the name of my constituency.

With what looked like a spider with a wooden leg had signed it, I returned his pen, he went to his chair, sat down, and with a smile, motioned me to take my seat. What the hell do I do now, I had to pretend I didn't see him, so I bowed my head, as my companion turned me round and we were on our way out of the chamber for the last time. As we rounded the back of the Speaker's chair, and off and out of the chamber, my wedding ring fell off and rolled down the corridor. How embarrassing was that, my companion managed to grab it before it rolled down a grating. Without a word from either of us he picked it up and put it in my hand. As my husband and the boys came down from the visitor's gallery, they gave me a big hug, then I introduced them to my colleague and left them a while as they chatted politely for a moment. I told my husband about my wedding ring falling off, all he could do was smile sympathetically and say he would have it made a bit smaller for me when we get home. I bought some smaller knickers but never gave a thought to my fingers getting thinner. After a few minutes, my husband had to nudge them as we turned to leave, and under my breath, thought the sooner the better, my nerves were beginning to get the better of me. It's a shame we have to go, I would have been thrilled to bits if we could have taken the boys for tea on the terrace, you know the one by the river, that you can see on the tele', but it was not to be , or ever likely to be. As we retraced our steps to the lift and out into the fresh air, we heaved a huge sigh of relief under our breath. As the nurse came to help me, we nearly had kittens on the spot! The party leader came rushing after us, what the hell does he want, but we needn't have worried he just wanted to wish me well and hoped to see me soon. We thanked him and said we hope so too.

The sun was shining as we came out of the shadows and into "New Palace Yard" I was beginning to feel a bit sickly and asked the nurse to hurry. Ten minutes to three, by now I couldn't wait to get home, the ambulance driver and nurse had been looked after very well by the security men with tea and a sandwich, they even bought us a sandwich each, you don't need me to tell you who ate them both. I didn't say anything to my husband, but I was feeling a lot worse than I let on as my stomach churned, and my chest felt tight. I just wanted to sleep all the way home. We thanked the security guards for their kindness, then we said goodbye to the boys as they went to find their car. They said it was easy along the M4 and they shared the cost of the petrol. Drive carefully I called out, as any mother would. We'll see you on Sunday afternoon, they shouted as they disappeared round the corner. The ambulance driver planned to stop once we are clear of London and top up with diesel. My husband's car would need some petrol too. I think they stopped once for the toilet, but I really don't know, and cared even less, as I said, I just wanted to get home and remember little of the journey. It seems the nurse gave me something to help me sleep.I don't know what time we got home, but I can just recall the feeling of having met my goal, to become a Member of Parliament for a second term. One thing I do remember, is seeing the stars as I was being wheeled into the house and wondered if Billy Bigalo was up there waiting for me.The following morning, my husband brought me my usual cup of tea and asked how I was feeling, a little tired I said, don't worry, I've called the doctor, he'll be here soon just to check you over after your day out. Following that long journey yesterday I felt dirty and would love to have a shower, but knew it was out of the question, so I asked for the ultimate in love between a man and his dying wife. I asked him to bathe me before the doctor arrived. Without a moments hesitation he said you know I will. A few minutes later, he came back to my bedside with a bowl of apple blossom scented warm water and a nice soft towel. As I sat on the edge of the bed he ran the warm scented cloth over my feeble body, I could see the tears welling up in his eyes, the only response I could make was to close mine and just feel his loving hands over me and try my best not to cry too. After a clean nightie and a cup of tea,I felt miles better. About ten thirty I heard the doctor let himself in and call up stairs, he stopped on the landing and spoke to my husband, then came in to see me and asked how I was feeling, not as bad as I expected I said, but my back still aches a bit.I went on to tell him I had been coughing during the night. He didn't say in quite so many words but I could sense a feeling of, I told you so. He gave me a shot of something to help me cope with the pain, and as he looked at me, he said a spell in hospital will help you to get over this bad patch. He spoke once more to my husband, then went down stairs to 'phone for an ambulance to take me to the local hospital as soon as reasonably possible.On what I thought was the next day I woke up in a side ward to see my husband sitting in a chair sipping a cup of tea, I was so glad to see him. He leaned over and held his tea cup to my lips. It was beautiful, as nectar is to a bee. He said you've been asleep all afternoon, what day is it. Its still Tuesday he said in his nice warm manner.

My back was still aching so I could only just get up on my elbows, It was so frustrating, I could hear people milling about, but couldn't see anyone.My chest still felt a bit tight, but not as bad as it has been.The effort of getting up on my elbows just to be nosey was to much for me, and I fell asleep once more. I woke to see my hubby fast asleep on a portable bed, rather than leave me. Then a nurse gave me a drink of something, I think it was tea, I drank it all down as I was so thirsty, and lay there, wondering what was to become of me. It's odd, but I have no fear of dying, as I thought I might have, when I got to this stage of my cancer. I just lay there watching a fly walking across the ceiling. The next morning, I was in another ambulance going to another hospital, or so I thought. But to my relief I could just about make out in the folds of the nurse's uniform the two words that are a great comfort to any cancer sufferer, "Macmillan nurse". I asked her where we were going. With a beautiful smile in her eyes and her lovely dark brown Caribbean voice she said we were going to the local Hospice my love, we'll look after you. She said your husband must love you very much to arrange this move for you at such short notice, he's gone home for a clean nightie and some bits and bobs for you. And I think he mentioned something about a blue dressing gown. Thank God I thought, for George Macmillan. You may wonder why I said thank God, well, I know the end is near, so what better place to spend my last few weeks before I shuffle off this mortal coil, than a hospice, with beautiful lawns and gardens, handsome doctors and the best nurses in the world. But most of all someone to talk to, I feel like I've been in isolation for this past few months, with all the deception and hiding from people all the time.Even through the haze of drugs I knew my husband was by my side. Where are my two boys I asked, I do so want to see them soon.

He said he had 'phoned them to let them know where I am, and went on to say they were quite concerned for me, being in a hospice, and will be here on Sunday early afternoon. I think today is Wednesday, but I slept most of the day. But the next day, as I lay there waiting for my husband, I felt a lot better, better in fact, than I have for a long time. About ten o'clock he came in, knelt by my bed and took my hand in his and put my wedding ring back on my finger, and wrapped the famous big blue dressing gown round my shoulders.I couldn't help but shed a tear as we held hands, and soon I began to feel a even better, and able to sit up in bed and look at my get well cards, both of them. As I felt better, I had a little wheelchair ride out into the garden and enjoyed a bit of sunshine with my husband. We sat in the sun for nearly two hours, it was lovely. We had a long talk about what we've just achieved and the ethics of the situation, not to mention the moral aspect. Now that it's all over and I'm an active Member of Parliament again, I feel, deep down that that was the real reason for my actions over the last eight months or so. Just be a Member of Parliament once more. I think I used the gratuity payout for my husband and the boys as a stick to beat myself with, and to excuse what I was doing. I think this might be a good time to tell you something. They say confession is good for the soul, well, here goes. These past twelve months have been hell, not just about my illness, but my conscience too. In the beginning, I mean the first time I had cancer, I was furious with Him or Her up there. But when I got used to the idea, had all the treatment, followed by a clean bill of health, I forgot my troubles and got on with life for a good few years. But when it struck for a second time last year, I was even more furious than before. Why me, twice! that's why I decided to do what I did. But now I am beginning to think I'm not proud of what I've done, with all the lies and deceit, involving my husband and our two sons in what is nothing less than a horrid con trick on my faithful supporters and helpers, and all the people who sent me money for my campaign. I behaved with greed and total lack of dignity, just to secure more money than we really need. Do I feel better for telling you, I think I damn well do. By now my appetite was a little better, maybe some chicken soup and a couple of slices of bread. I couldn't help but think the food bill here must be minimal, but the nurses are absolutely without equal. Friday, I sat up again, but didn't go outside, it was raining. We sat in the lounge with two other ladies but didn't say more than good afternoon and a few polite words. After my lunch, I tried to read a little but I couldn't concentrate or see the point. During my time here my husband has been by my side most of the time, now I'm a little better, I made him go home and have a good nights sleep, Oh! and have a shower. As dawn broke on Saturday, I heard someone say, her husband will be pleased as I woke up to the warm feeling of my husband's hands holding mine with his firm reassuring grip, his very touch says he loves me so much. He was more than pleased to see me looking so much better. My "resurrection" was the cue for my him to call the boys with the good news. But I was still very tired and as I drifted in and out of sleep, I had a dream, could it be Him,or Her, up there, I think it must be a woman,did my dad really open the pearly gates for mum, and was that really Billy Bigalo polishing the stars. Now that I have told you why we did what we did, I feel I cannot, as I said earlier, see it through to the bitter end. I had another long talk with my husband and even after all the scheming and the worrying, there was only one honourable course of action to save my reputation, and die with dignity. I 'm going to apply for the Chiltern Hundreds as soon as possible, and resign my seat. Yes it will be a shame to give up all that money. But I do have a good portcullis pension fund.To stay on as an active Member of Parliament till I die, well what were we thinking of. It was hard work on Monday going down to Westminster just the once, let alone as a member on a regular basis. It would have been a sure way to bring shame down on my husband and our two boys if anyone was to cry foul. I'll be happy to go with more dignity than I deserve, with a job well done, as a Member of Parliament, a loving wife and mother, in a very happy marriage, with two wonderful sons, and this feeling of relief that I have managed to put right some of the wrong I have done. With a cheque for £10,000 given back to the local constituency, being close to the amount spent on my campaign and the most grovelling apology imaginable to head office. After all, money Isn't everything, the doctors at the hospice said I could go home next week if I felt up to it, try and stop me I thought. So, with arrangements made for me to go home next week, and the boys soon to take their summer holiday, we can all look forward to a few weeks or even months together in the summer sunshine and maybe an apple from my Malus Domestica.

About the Author

Thou' not a literary masterpiece. This brand new author as come up with an original story based on life's experiences and just how scheming, people can be when presented with stark choices. She is so easy to read you won't be able to put this book down till you have read it. Then you will want to read it again just to be sure.

Lightning Source UK Ltd.
Milton Keynes UK
30 December 2009

148010UK00001B/30/P